Anguel S. Stefanov
Gregorie Dupuis-Mc Donald
Editors

I0030951

Spacetime Conference 2022

Selected peer-reviewed papers presented at the Sixth International Conference on the Nature and Ontology of Spacetime, 12 - 15 September 2022, Albena, Bulgaria

MINKOWSKI
Institute Press

Anguel S. Stefanov
Bulgarian Academy of Sciences
Str. "15 November" 1
1000 Sofia, Bulgaria

Gregorie Dupuis-Mc Donald
Department of Philosophy
University of Salzburg
Franziskanergasse 1
5020 Salzburg, Austria

Front Cover: Most participants of the Sixth Spacetime Conference. Picture taken by Sho Fujita

ISBN: 978-1-989970-96-6 (softcover)
ISBN: 978-1-989970-97-3 (ebook)

Minkowski Institute Press
Montreal, Quebec, Canada
http://minkowskiinstitute.org/mip/

For information on all Minkowski Institute Press publications visit our website at http://minkowskiinstitute.org/mip/books/

Most participants of the *Sixth Spacetime Conference* held in the famous resort Albena (near Varna) on the Bulgarian Black Sea coast from 12 to 15 September 2022.

Front row (from left to right): Georgios Kallidis, Hiroaki Fujimori, Svetla Petkova, William Giovanni Jimenez Senzano, Anguel S. Stefanov, Gregorie Dupuis-Mc Donald, Sho Fujita, Jerzy Gołosz

Back row (from left to right): Marko Vojinovic, Marcoen Cabbolet, Gall Alster, Vesselin Petkov, Robert Rynasiewicz, Dan Shanahan, Bruce M. Boman, Maire Shanahan, Branko Kovac

CONTENTS

Foreword by the Editors   vii

Contributors   xi

I   PHILOSOPHY AND PHYSICS OF SPACETIME   1

1   What is essential to constitute a physical realm? Hints from a spatiotemporal structure
Sho Fujita   3

   1   Introduction: Contemporary Physics Deals with Physical Realms Abstractly 3
   2   Abstract and Concrete Properties of Spacetime 6
      2.1   Between "Theory" and "Actual World" 6
      2.2   Privileged Spacetime Realm 9
      2.3   What Does Spacetime Refer to? 12
   3   Extended Physical Realms 15
      3.1   Give-and-Take Relation Between Topology and Metric 16
      3.2   More Abstract in Micro Regions 22
   4   Conclusion: A New Way of Interpretations About Spacetime Through Macro and Micro Region 26

2   Imagination, Fiction and the Reality of Minkowski's Discovery of Spacetime
Gregorie Dupuis-Mc Donald   33

   1   Introduction 33
   2   Imagination, reality, and scientific discovery. An overview of the problem 35
   3   The philosophical debate on the reality of spacetime . . 38
   4   The role of explanation for Minkowski's discovery 40
   5   The role of imagination in the discovery of spacetime . . 42

6   How Minkowski's discovery can be taken as the outcome of a fictionalist strategy 45

7   Conclusion 50

3 Einstein and Wittgenstein's ladder
Elton Marques   53

1   Introduction 53

2   Relativity of simultaneity and divergent voices 55

     2.1   *RoS* as a conventional result 56

3   *RoS* as a robust result 59

     3.1   The supposed equivalence between interpretations 62

     3.2   The definition of simultaneity 67

4   Conclusion 71

II   THE ONTOLOGICAL NATURE OF SPACETIME   77

4 Is Spacetime Emergent?
Anguel S. Stefanov   79

1   Short Introduction 79

2   What is it for Spacetime to be Emergent, and not a Fundamental Entity? 80

3   Functionalism 82

4   A Remark about the Posited Existence of Non-Spatiotemporal Grains of Matter 85

5   My Answer to the Title Question 87

5 Experiment: Mass does increase with velocity
Vesselin Petkov   91

1   Introduction 91

2   Relativistic mass is an experimental fact 97

3   Addressing objections against relativistic mass 102

4   Conclusion 106

6 Dynamic Multipresentism: In Defence of a Dynamic
View of Reality
Jerzy Gołosz   113

1   Introduction: The Problem with Dynamics and the Flow of Time 113

2   Physics and the Flow of Time 115

3   Philosophy and the Flow of Time 117

4   Final Remarks: Metaphysics as a Base for Physics? . . . 122

III INTERPRETATIONS OF RELATIVITY AND GRAVITATION 127

7 The nature of inertia, and the dynamic gravitational field
Branko Kovac 129
  1 Introduction 129
  2 Estimating the strength of the dynamic gravitational field 133
  3 The equation for the dynamic gravitational field 134
  4 The motion of the planets and free fall 138
  5 Design of the experiment to detect the dynamic gravitational field 141
  6 Properties of space around the moving masses 143
  7 Dark matter or dynamic gravitational field 147
  8 Conclusion 149

8 Protogravity: a quantum-theoretic precursor to gravity
Daniel Shanahan 153
  1 Introduction 153
  2 The twin effect 156
  3 Newtonian gravity 160
  4 The Schwarzschild metric 163
  5 The metric components g_{tt} and g_{rr} 165
  6 The de Broglie wave 169
  7 Quantum mechanics from particle wave structure 172
  8 Gravity from particle wave structure 176
  9 Concluding remarks 178

9 On the special principle of relativity
Hiroaki Fujimori 183
  1 Preliminaries 183
    1.1 Terms 183
    1.2 Special principle of relativity 184
    1.3 Theorems 185
  2 Introduction 186
  3 Proof of the special principle of relativity 189
    3.1 Symmetry plane 189
    3.2 What k gives? Three types of a symmetry plane 194
    3.3 Symmetric spacetime structure of two inertial coordinate systems 197
  4 Conclusion 200

10 Noncompactified Kaluza-Klein Theories
S. M. M. Rasouli   **203**

  1   Introduction 203
  2   Five-dimensional Ricci–Flat Space and the IMT 205
  3   Modified Brans-Dicke Theory 207
  4   Exact BD-KS anisotropic vacuum solutions
        in five-dimensions 210
  5   Effective BD-KS cosmology on a four dimensional hypersurface 212
  6   Conclusions and discussions 217

FOREWORD BY THE EDITORS

This volume contains a collection of peer-reviewed essays which presents contributions that were made at the sixth session of the international conference "Nature and Ontology of Spacetime", organized by the Minkowski Institute, and which was carried out in Albena, Bulgaria, 12-15 September 2022. That conference brought philosophers and physicists together, and created a forum for fruitful debates concerning the nature and ontology of spacetime. Topics in relativity theory, quantum mechanics, cosmology and philosophy of science were discussed.

This volume proposes discussions and solutions of various physical, ontological and epistemic problems in contemporary philosophy and history of spacetime physics, spacetime and quantum gravitation, the nature of time, relativistic phenomena and the reality of spacetime ontology, as well as aspects of the mathematical structure of special and general relativity theory. We hope the reader will appreciate the broad spectrum of topics and questions covered, and will be entertained by the ideas presented by our contributors.

The volume is organized thematically, and contains three sections. The first section considers aspects of the PHILOSOPHY AND PHYSICS OF SPACETIME. Sho Fujita deals with the elucidation of the philosophical requirements needed for the constitution of a spatiotemporal physical realm. Gregorie Dupuis-Mc Donald considers the nature and status of the scientific discovery of Minkowski spacetime. Elton Marques considers the importance of empiricism and realism in understanding the debate on whether simultaneity is absolute or relative.

The second section contains essays on THE ONTOLOGICAL NATURE OF SPACETIME. Anguel Stefanov studies whether spacetime is an emergent entity, a conviction that is at the heart of theories of quantum gravity. Vesselin Petkov argues that relativistic mass is an experimental fact, and shows why the rejection of the relativistic velocity dependence of mass amounts to a rejection of experimental facts. Jerzy Gołosz essay in the philosophy of time defends dynamic multipresentism, an argument for the existence of the flow of time.

The third section presents INTERPRETATIONS OF RELATIVITY AND GRAVITATION. Branko Kovac studies the nature of inertia and the dynamic gravitational field and examines the hypothesis that the inertial force develops because the accelerating mass creates a force field around it in the same way as the weight of a mass is a consequence of the Newtonian gravitational field. Dan Shanahan studies protogravity as a binding effect, and argues that it was sufficient in itself to explain the preference of matter for bound, rather than free, motion in the early universe. Hiroaki Fujimori presents an original discussion of the Relativity Principle, and shows what are the implications of Euclidean geometry remaining invariant and holding completely on both the front and back symmetric plane? Finally, Mehrdad Rasouli presents a review of noncompactified Kaluza–Klein theories and together with their motives and implications for cosmology.

Acknowledgements

As editors of this volume, it is a pleasure to recognize Minkowski Institute's academic mission, and to thank Vesselin and Svetla Petkov for their continuous and dedicated work contributing to the dissemination of scientific knowledge. Also, we want to thank all participants at the conference for their valuable contributions and hard work.

Anguel S. Stefanov and Gregorie Dupuis-Mc Donald, the editors

Contributors

Gregorie Dupuis-Mc Donald
Department of Philosophy
University of Salzburg
Franziskanergasse 1
5020 Salzburg, Austria
gregorie.dupuis-mc-donald@stud.sbg.ac.at

Hiroaki Fujimori
Tokyo, Japan
fuji@spatim.sakura.ne.jp

Sho Fujita
Department of Complex Systems Science
Nagoya University Graduate School of Informatics
Furo-cho Chikusa-ward, Nagoya-City
464-8601, Aichi, Japan
fujita.sho.p6@a.mail.nagoya-u.ac.jp

Jerzy Gołosz
Institute of Philosophy
Jagiellonian University
ul. Grodzka 52, 31-044 Kraków, Poland
jerzy.golosz@uj.edu.pl

Branko Kovac
Sydney, Australia

Elton Marques
Universidade Católica Portuguesa
elmarques@ucp.pt

Vesselin Petkov
Minkowski Institute
Montreal, Quebec, Canada
http://minkowskiinstitute.org/
vpetkov@minkowskiinstitute.org

S. M. M. Rasouli
(1) Departamento de Física
Centro de Matemática e Aplicações (CMA-UBI)
Universidade da Beira Interior
Rua Marquês d'Avila
e Bolama, 6200-001 Covilhã, Portugal
(2) Department of Physics
Qazvin Branch, Islamic Azad University
Qazvin 341851416, Iran
mrasouli@ubi.pt

Daniel Shanahan
Brisbane, Australia
danjune@bigpond.net.au

Anguel S. Stefanov
Bulgarian Academy of Sciences
Str. "15 November" 1
1000 Sofia, Bulgaria
angstefanov@abv.bg

Part I

PHILOSOPHY AND PHYSICS OF SPACETIME

A. S. Stefanov, G. Dupuis-Mc Donald (Eds), *Spacetime Conference - 2022.*
Selected peer-reviewed papers presented at the Sixth International Conference on
the Nature and Ontology of Spacetime, 12 - 15 September 2022, Albena, Bulgaria
(Minkowski Institute Press, Montreal 2023). ISBN 978-1-989970-96-6 (softcover),
ISBN 978-1-989970-97-3 (ebook).

1 WHAT IS ESSENTIAL TO CONSTITUTE A PHYSICAL REALM? HINTS FROM A SPATIOTEMPORAL STRUCTURE

SHO FUJITA

Abstract   Space and time are fundamental factors in describing physical phenomena. Metrical features such as distance and angle are necessary and sufficient to describe physical spacetime. Contemporary philosophy of spacetime regards spacetime as a metric field. Structural realism applied to spacetime supports this interpretation, viewing a spatiotemporal structure as concrete in the physical world. In General Relativity, the spatiotemporal structure has mathematical substructures at different stages, with abstract structures like topology being defined mathematically without metric. Although topology is independent of locality and can *arise from* the metric, it may be essential for physical reality in quantum gravity theories. Abstract features are more important for physics as a spatiotemporal structure given by metric emerges from more fundamental entities in micro regions.

Keywords: Structural realism, identity, hole argument, FLRW metric, emergence

1   Introduction: Contemporary Physics Deals with Physical Realms Abstractly

When pondering the ontological question of the reality of space and time, the focus is usually on physical spacetime. Both physics and philosophy have primarily been studying this concept of spacetime, initially exploring it as the physical realm in which concrete entities like ourselves exist.

According to physics, spacetime is described as being curved by Riemannian non-Euclidean geometry in the context of the General Theory of Relativity (GTR). Contemporary philosophy of spacetime has also

A. S. Stefanov, G. Dupuis-Mc Donald (Eds), *Spacetime Conference - 2022.*
Selected peer-reviewed papers presented at the Sixth International Conference on the Nature and Ontology of Spacetime, 12 - 15 September 2022, Albena, Bulgaria (Minkowski Institute Press, Montreal 2023). ISBN 978-1-989970-96-6 (softcover), ISBN 978-1-989970-97-3 (ebook).

considered what can be considered as spacetime. In GTR, spacetime is described by a complex mathematical structure known as a manifold, which is a global topological space that includes relational local features such as connection and metric. Philosophy of spacetime must deal with abstract entities. In fact, the relationship between mathematical models and physical spacetime is a key issue for philosophy to address. This demonstrates how mathematical objects and structures exist in the physical world.

Recently, the concept of space has become broader and is influencing the ontology of spacetime. In mathematics, there are other types of abstract space, such as topological space and vector space, which are derived from sets where each element is constituted by specific relations. Additionally, physical worlds are now understood to be more extensive than in the past. In contemporary physics, theories such as GTR and string theory introduce high-dimensional realms beyond four-dimensional spacetime, and quantum theory uses phase spaces and configuration spaces to describe N-particle systems. These types of space play important roles in explaining various physical phenomena, including micro regions, but it is important to note that their roles are distinct from that of spacetime.

Simply put, Hilbert space and Fock space in quantum theory are not real as physical entities. This is because these spaces are not physical realms, but rather useful tools for explaining physical observables in quantum phenomena. They are used by physical theories, not by mathematical theories. Nevertheless, we cannot locate these spaces anywhere in the world.

But the situation is more complex. Paul argues that these tool-like spaces are more fundamental than spacetime (Paul 2012). According to him, the 3-N dimensional configuration space occupied by N-particle systems is the true reality, and spacetime is not a fundamental constituent of this physical world. In some quantum gravity theories, where micro regions are dominated by quantum effects, spacetime can be seen as emergent or derived from more fundamental entities (Wüthrich 2012, 2018; Huggett & Wüthrich 2013), such as causal sets or spin networks, which I will discuss in Section 3. This emergent nature of spacetime presents a new worldview that challenges the traditional metaphysics of spatiotemporalism.

Therefore, the above ontological question encompasses various interpretations, leading to numerous forms of realism and anti-realism about spacetime (Slowik 2015). However, I aim to clarify the mathematical concept that plays a crucial role in determining physical spacetime and to demonstrate what constitutes a physical realm. This clarification is directly linked to the traditional debate about realism in spacetime, namely substantivalism versus relationism, which has its roots in the historical discussions between Newton and Leibnitz about spacetime as a container for matter.

Furthermore, when it comes to spacetime in GTR, a structural interpretation (Esfeld & Lam 2008; Lam & Esfeld 2012) can be very useful in understanding how a mathematical structure corresponds to a physical one. This correspondence relation (Psillos 2010, 2011; Pincock 2007; Suppe 1989) and discussions about contemporary philosophy of spacetime (Earman & Norton 1987; Maudlin 1988; Teller 1991; Hoefer 1996; Dorato 2000) will be addressed in section 2. The main point of discussion in GTR is the interpretation of the gravitational field (2.2 and 2.3). In mainstream discussions, the gravitational field is considered to be spacetime as it is also a metric field in GTR and expresses local properties of spacetime as spatiotemporal features. This view will be explored further.

From the possible interpretation of macro spacetime, I aim to explore the aspect of spatiotemporal structure that is retained in more fundamental entities in micro regions where spacetime emerges. Some quantum gravity theories abandon ordinary properties of spacetime such as distance and angle, and quantizing gravity according to these theories may lead to the conclusion that spacetime does not exist in micro regions. This is because if spacetime is comprised of gravitational fields, then quantizing gravity would mean quantizing spacetime itself. However, even if spacetime transforms into different entities through quantization, I believe these new entities must still reflect some aspects of spacetime, as long as their structures are defined by quantizing something from the spatiotemporal structure.

There must be a physical realm, no matter how small, in micro regions. This realm may be very different from ordinary spacetime, but the abstract mathematical substructure of the entire spatiotemporal structure is what is essential for constituting a physical realm, including spacetime. This raises the question again, what makes spacetime what it is? In section 3, I will demonstrate that not only local, but also global properties are necessary to identify spacetime points, even in macro GTR, using the symmetric universe model as an example. This suggests that abstract properties, rather than concrete ones such as metric, are essential for being a physical realm.

2   Abstract and Concrete Properties of Space-time

In this section, I will examine the current state of philosophy of space-time. To understand the existence of spacetime in the context of GTR, it is necessary to examine the relationship between physical spacetime and the mathematical models used to describe it. First, I will look at how abstract entities in scientific theories explain events in physical worlds.

2.1   Between "Theory" and "Actual World"

In philosophy of science, there is a question of the existence of theoretical objects as claimed by scientific realism. This relates to the interpretation of scientific theories and the correspondence between models and actual worlds. For example, Frigg points out that models share important aspects in common with literary fiction or pretence and give a general picture of scientific modelling (Frigg 2010). Surely theories and models may not accurately represent an actual world due to approximations or idealizations (Cartwright 1983). The phrase "theoretical objects such as electrons and mass points exist in our actual world" may not be accurate.

Psillos argues that theoretical objects such as electrons and mass points exist in models as universals, and that scientific realism that presupposes abstract theoretical objects leans towards Platonism rather than nominalism (Psillos 2010, 2011). According to Psillos, mass points described in physical theories can represent individual massive bodies in the actual world, but they do not directly refer to these physical bodies. He states that models "are not concrete and...not causally efficacious" (Psillos 2011, 4), but that they still have explanatory power.

> The processes of idealization and abstraction are such that the description of the model isolates the explanatorily relevant features of the represented system with respect to the behaviour under study. It specifies the basic or more central explanatory mechanism or regularity. (Psillos 2011, 16)

Theoretical objects and structures as abstract universals correspond to something concrete. The Linear Harmonic Oscillator (LHO), for example, can accurately explain various calculated values of the swing of specific pendulums. This interpretation can also be applied to unobservable entities, such as electrons, which are discussed within the context of scientific realism versus antirealism. Unlike macro bodies, electrons cannot be directly observed, but their motion can be influenced and intervened with, according to Hacking (1983). Even if the concrete entities that instantiate or exemplify the theoretical electron are different from those electrons, there must be a causal entity in the actual world for theories to be verified through experiments and observations. In this sense, scientific realism presupposes that specific electrons or their counterparts exist in the actual world.

Models of data and structures are described mathematically in theories, but they also exist in the actual world through a correspondence relationship, as Psillos acknowledges.

> For now, the question is whether the model of the data itself (let us fix our attention on this to make things easier) is n-adequate [nominalistic adequate] vis-a-vis the phenomena, and answering this question presupposes either a direct confrontation of the model with the (unstructured) phenomena or the comparison of the model with another—one that (presumably) captures the causal structure of the phenomena. The first option does not seem to make much sense. The second option requires that the phenomena (or the world) have a built-in causal structure. (Psillos 2010, 956)

This interpretation suggests that theoretical entities and structures in models and theories correspond to structures in the actual world, based on isomorphism, and so on. From the perspective of limited scientific realism, models only need to capture certain aspects of the world, as Giere points out.

> We can also imagine that the two models are equally endowed with any supposed superempirical virtues such as simplicity or unity. Here I am strongly inclined to say that there can be no scientific basis for claiming that one model better fits the overall structure of the universe. Again, we have a limit on realist claims. (Giere 2004, 751)

If we want to adopt a nominalistic position, "[i]t is a further and separate claim that the model of the data (or the theoretical model for that matter) adequately represents concrete (causal) physical systems (or patterns). For the theory to be n-adequate, it is the latter claim that has to be true" (Psillos 2010, 956). Ketland says that structures are "nominalistically equivalent iff their concrete parts are isomorphic"

(Ketland 2010, 208) meaning that the concreta behave 'as if' the theory is true.

It is important to interpret abstract entities or structures introduced in theories correctly. Nominalists might eliminate or reduce them, partly influenced by Field's belief that there is no abstract entity (Field 1980). Pincock's fictionalist view holds that mathematical systems are essential for theorizing physical worlds ("theoretical indispensability"), but not necessary for determining what exists ("metaphysical dispensability"). According to Pincock, when doing physical science, entities or structures in mathematical systems provide us with accurate knowledge, but this knowledge should be supported by empirical evidence, not theoretical content. Only with physical facts can these existences and properties in a theory be verified, and pure mathematical content in the theory can be disregarded.

> it is coherent and sensible to maintain that *the actual bottom-level physical facts* render the nominalistic content of empirical science true and the platonistic content of empirical science is fictional. (Pincock 2007, 268[1])

As an example in his paper (2007, 267), if a temperature theory (T) is incomplete and the physical basis for temperature is still unresolved, with T leaving the crucial interpretive question of the existence of the lowest temperature open, facts about whether temperature has a lowest value should be determined based on the empirical evidence available to us. Not until we gain evidence, will the lowest temperature come into existence and its theoretical counterpart will be eliminated.

This eliminative approach supports qualitative parsimony. Adopting Pincock's perspective means we don't have to accept the existence of an excess of theoretical structures that are mathematically described. In this case, models are merely epistemological tools used to super-empirically interpret the world, rather than ontological entities (Suppe 1989). In addition, if we were to separate the causal structures from other theoretical elements that are not supported by empirical evidence, we could avoid making reference to others (van Fraassen 1980, 2006). Theoretical abstract entities have a different status from that of physical entities.

What is a causal structure? Besides the argument between realism and nominalism, if we accept the distinction between a model or language-like entity in scientific theories and the actual physical world,

---

[1]His refined nominalistic position is devised from Balaguer's approach (Balaguer 1998) to challenge Quine's ontological commitment.

the latter refers to a causal world in which phenomena occur and can be directly experienced through touch and observation.

That is to say, the causal world is an actual physical world where particular objects and structures exist. This concrete realm differs from abstract ones and may refer to a spacetime realm, namely our universe[2].

Surely, the world constituted by spacetime has special features. At least, a spacetime realm is a part of the physical world, whether it is our world or another parallel world, as long as it is not in fiction, for example, novels. Philosophy of science, which deals with the realism of unobservable micro physical entities such as electrons, also connects causality strictly to a spacetime realm. Chakravartty establishes criteria of realism as entities having causal effects through detections (Chakravartty 1998, 2007), namely in the spacetime realm. In the next subsection, I will concentrate on this spacetime realm and traditional discussions of spacetime itself.

2.2 Privileged Spacetime Realm

It is true that many of philosophers and physicists consider spacetime to be the clearest realm. As mentioned in the previous subsection, physics originally deals with natural phenomena that occur in spacetime where matter exists with a particular position referred to by four-dimensional coordinate values (t, x, y, z). Philosophy, on the other hand, has addressed the realism of not only matter but also mind, mathematical numbers and sets, which do not exist in spacetime. The problem of universals is about whether these abstract entities exist in realms other than spacetime. In essence, do physicists focus on matter and spacetime while philosophers of physics, mathematics, and metaphysicians focus on abstract entities and realms?

But especially in contemporary physics since the 20th century, physical phenomena have been extended beyond spacetime. GTR presupposes a high-dimensional spacetime as a new physical realm in which black holes exist. In addition, quantum gravity theories and quantum cosmology are developing more fundamental entities from which spacetime derives or emerges, namely the origin of spacetime itself. Of course, physics before the 20th century also used abstract concepts such as "action" and "Lagrangian" described in analytical mechanics, but their mathematical behaviors based on fundamental laws

---

[2]Accurately, the realm we can interact causally with is limited to only within the light cone emitted from us.

are merely theoretical explanations for phenomena in the spacetime realm, rather than physical phenomena themselves. At least in the past, physical phenomena were limited to the spacetime realm.

Spacetime ontology has been a topic of discussion for centuries, but it is only relatively recently that structural interpretations have been incorporated into the philosophy of spacetime based on GTR. These new interpretations have partially influenced the traditional debate about the realism of spacetime, substantivalism and relationism, which dates back to the Newtonian era.

In Newtonian physics, spacetime can be considered as a background for matter. It is assumed that spacetime acts as a container for matter. However, a key question still remains: Does spacetime exist independently of matter? This question deals with whether or not the container, made up of space and time, remains even in the absence of matter. This raises the issue of the realism of empty spacetime, or a series of momentary vacuums. From a metaphysical perspective, this empty container is considered as absolute spacetime, which according to Newton, requires God's perception of matter (Newton, based on Alexander 1956, p.15), representing an early substantivalist interpretation.

Spacetime consists of infinite spacetime points, and each point may be considered real according to the substantivalist viewpoint in the context of Newtonian mechanics. These spacetime points are mathematically represented as elements of a set that spans the entire spacetime. It is still possible for substantivalists to hold a belief that spacetime points don't have intrinsic properties, despite the fact that such a belief supports Leibniz-shift being an effective counterargument against substantivalism. This viewpoint is in line with Newton's own belief, as stated below.

> For just as the parts of duration derive their individuality from their order, so that (for example) if yesterday could change places with today and become the later of the two, it would lose its individuality and would no longer be yesterday, but today; so the parts of space derive their character from their positions, so that if any two could change their positions, they would change their character at the same time and each would be converted numerically into the other. The parts of duration and space are only understood to be the same as they really are because of their mutual order and position, nor do they have any hint of individuality apart from that order and position which consequently cannot be altered. (Newton 1962, p.136)

That is to say, spacetime points should be identified based on their specific positions in the whole. Even in Newtonian physics, a spacetime realist like Newton can avoid presupposing haecceities (I will discuss this further in 3.1) for spacetime points, instead opting for structural interpretations. Spacetime points are geometric elements that are a priori indistinguishable from each other, not algebraic elements that can be distinguished from each other based on their properties (Stachel 2002).

This worldview also extends to spacetime in GTR. The 20th century field theory introduced a groundbreaking perspective, where spacetime is non-flat and has a non-Euclidean geometry described by metric field functions. In GTR, these fields are considered physical gravitational ones, and the spacetime realm mathematically consists of infinite spacetime points. Despite the shift in spacetime theories from Newtonian physics to Einstein's GTR, the interpretation of spacetime points remains similar.

In GTR, geometrically each point has its own properties, including metrical relations with neighbouring points defined by a metric tensor g_{ik} throughout the spacetime realm. The metric fields give all spacetime points spatiotemporal relational properties, and they locally describe how a specific realm of spacetime is curved.

> However, the metric tensor at a space–time point cannot strictly speaking be understood as an intrinsic property, since it involves infinitesimally neighbouring space–time points through the notion of tangent space on which it is defined. (...) [T]he fundamental space–time properties would be relational only in an infinitesimal sense so that the fundamental relations are only infinitesimal relations (Lam & Esfeld 2012, 248-249)

Metric plays a crucial role in establishing the relational properties of each spacetime point.

In recent discussions of the philosophy of spacetime, both substantivalism and relationism widely acknowledge that spacetime should be seen as a metric field (Slowik 2004, 2015). The former argues that this dynamic spacetime exists independently of other physical fields, such as metric field substantivalism (Hoefer 1996), while the latter claims it is just a property of or can be reduced to other physical fields, like dispositions (Teller 1991). In light of this ambiguous realism of metric fields, the conventional distinction between substantivalism and relationism has become outdated (Rynasiewicz 1996) and structural realism about spacetime is considered a third perspective (Dorato 2000). All of these discussions are based on the assumption that the essence of physical spacetime is the metric field.

For ontological structural realism (OSR), especially for the moderate versions proposed by Esfeld and Lam, this picture of spacetime points is very consistent, even though the spacetime in Newtonian physics differs radically from that in GTR. They view spacetime as "a mind-independent physical structure whose basic constituents have no fundamental intrinsic properties independently of the structure they are part of" (Esfeld & Lam 2008, 44). They regard these constituents as spacetime points at least in classical GTR, and they maintain that spacetime can exist independently, together with its relational properties, which are not reducible to the properties and relations of matter. "Moderate structural realism claims that the spacetime structure exists as a mind-independent physical network of spatiotemporal relations among spatiotemporal constituents (such as spacetime points) that do not possess any intrinsic properties" (Esfeld & Lam 2008, 42-43).

However, these structural interpretations make the realism of spacetime more complicated. In the next subsection, I will examine what these structural interpretations mean.

2.3   What Does Spacetime Refer to?

Of course, there are geometric properties other than metric about spacetime. In particular, manifold substantivalism claims that gravitational fields are kinds of matter and that spacetime points should have intrinsic properties as a manifold or primitive identities, namely haecceity, independent of relational ones given by metric (Hoefer 1996). I think that this idea, called manifold substantivalism, is strongly connected with a traditional conviction that spacetime is a container and exists independently of physical fields.

However, manifold substantivalism faces a serious problem. If we stick to the difference between spacetime and matter even in GTR, metric (gravitational) field may be included in the matter side for a manifold substantivalist. This view implies that a spacetime realm can exist without the metric field. This view leads to fatal indeterminism, namely the contemporary hole argument (Earman & Norton 1987).

The hole argument is concerned with the question of what orbit test particles move on inside H, a small hole with no other fields in a manifold M. Suppose a particle moves in M passing through H, and there are some orbit models in which diffeomorphisms from one to another manifold point are applied to a domain only inside H. In the same hole H, the particle passes through mathematically different points by diffeomorphisms. Therefore, only inside H, it seems as if there were different orbits and metric tensors in each model. If these apparently different models show physically different situations, geodesic solutions of the particle and g_{ik} can be obtained infinitely. This induces indeterminism for the geodesic equation and the Einstein equation.

Of course there are only passive coordinate transformations to realize diffeomorphisms based on general covariance in GTR between these models and we cannot observe a bare point without metrical properties. In fact, in order to observe which point the particle passes through, we use these properties, for example, through a ruler and a clock. What coordinate value given to a point depends on these items. That is to say, apparently different orbits are identified observationally with metrical information.

The metric field is necessary for identifying spacetime points. Earman and Norton argue that substantivalists "must either (a) accept that there are distinct states of affairs which are observationally indistinguishable, or (b) deny their substantivalism" (Earman & Norton 1987, 522). If we believe that a bare point in M has an identity independent of physical fields, including the metric field, as manifold substantivalism suggests, we cannot avoid fatal results. Contemporary substantivalism and relationism have developed by emphasizing the importance of the metric field in this contemporary version of the hole argument.

Apparently different metric fields diffeomorphically related as a group refer to a spatiotemporal structure. Stachel says that spacetime points in a manifold inherit all their chronogeometric (and inertiogravitational) properties and relations from metric fields. With that assumption, an entire equivalence class of diffeomorphically related mathematical solutions represents just one physical solution (Stachel 2002, p.233). Each solution is a different form to describe one common structure.

Intuitively, a structure absent from metric is so abstract that it runs short of physical spatiotemporal concepts such as distance, angle, and causality of light cones. In fact, it is very strange to call it spacetime. Einstein says as below.

> There can be no space nor any part of space without gravitational potentials; for these confer upon space its metrical qualities, without which it cannot be imagined at all. (Einstein 2007, p. 618)

Spacetime cannot be detached from a gravitational field in it and spacetime cannot be accounted independently of the gravitational field. Einstein himself explains this worldview as this.

> If we imagine the gravitational field, i.e. the functions g_{ik} to be removed, there does not remain a space of the type (I) [Minkowski spacetime], but absolutely nothing, and also no "topological space". For the functions g_{ik} describe not only the field, but at the same time the topological and metrical structural properties of the manifold. (...) There is no such thing as an empty space, i.e. a space without field. Space-time does not claim existence on its own, but only as a structural quality of the field. (Einstein 1961, pp.155-156)

This dynamical field interacts with others via the Einstein equation, and so spacetime is no longer a passive background. Before I conclude that metric field substantivalism is more refined than manifold substantivalism, I want to emphasize two points here as follows:

1. The worldview that "spacetime is a structural quality of the field" is another aspect of structural interpretations different than the discussion continuing from the previous subsection, namely how spacetime points are identified.

2. g_{ik} describes not only metrical properties but also topological properties of the manifold.

Spacetime and matter may be just different aspects of one structure. Regarding point 1, the question of what spacetime is becomes blurred and the division between spacetime and matter becomes more and more ambiguous, leading to super-substantivalism, which claims that spacetime refers to the whole universe, including not only the metric but also other material entities. This interpretation is influenced by Einstein's intention to unite gravity and the electromagnetic field.

Locality given by the metric field tensor leads to topology. As for point 2, I will focus on it in this paper. Einstein acknowledged that without the metric, it is impossible to describe not only physical spacetime itself, but also a topological space. It is difficult to imagine what would occur in the universe without gravitational fields (Maudlin 1990).

Is topology included in the metric? Hoefer argues that spacetime or part of spacetime can be described by the metric field, not the global topology (Hoefer 1996, 24-25). Maudlin claims that "the topology flows from the metric rather than the metric being imposed on the topological space" (Maudlin 1990, 554). Dorato also emphasizes that the topology of spacetime cannot be determined prior to the Einstein equation (Dorato 2000, 1610). I would like to suggest that topology, which is claimed by the three of them, is not a property previously possessed by spacetime, but rather emerges from all local metrical information. I think this is a doctrine of contemporary philosophy of spacetime.

3 Extended Physical Realms

While metrical features play an important role for spacetime and a spacetime realm, there is also a gap between metric and topology. A topological space is already a differentiable manifold consisting of continuous points and on which ordinary vector and tensor fields are defined.[3].

I wonder whether topology is included in metric in 2.3 or not, but mathematically, topology without metric is possible. In other words, even without local detailed properties, physical theories can be described to some extent with only global properties. Fundamental laws can be defined, and geodesic equations can be written without a metric because even the covariant differentiation of contravariant, covariant, and mixed tensors to establish differential equations can be defined on the condition that the concept of connection is introduced:

$$v^\mu \nabla_\mu v^\lambda = \frac{d^2 x^\lambda}{dt^2} + \Gamma^\lambda_{\mu\nu} \frac{dx^\mu}{dt} \frac{dx^\nu}{dt}. \tag{1}$$

$\Gamma^\lambda_{\mu\nu}$ is the Christoffel symbol, and this equation shows the parallel transport of a body for all orbits in a manifold or a topological space, leaving various possibilities about how the velocity of the particle is kept in moving along a geodesic line. Of course, although written formally, a connection without a metric is so abstract that we cannot specify the physical significance of geodesic equations.

This equation is an abstract geodesic equation. In addition, Maxwell's equations of electromagnetism can also be partially given without a metric. That is to say, abstract spatiotemporal properties can be described without a metric, and therefore, we should be careful about what is essential for physical spacetime or a spacetime realm.

---

[3]For details, refer to Hawking & Ellis (1975).

3.1 Give-and-Take Relation Between Topology and Metric

In this subsection, I suggest that the hole argument does not imply that the spacetime manifold is not physical spacetime. Sophisticated substantivalism and relationism argue that in order to describe our ordinary spatiotemporal concepts such as distance and angle, there must be a metric. I do not intend to challenge this worldview, but does the hole argument deny properties as a manifold by determinism?

In the discussion of the hole argument, the core idea for holding determinism is that the points inside H in a manifold are not individuated independently of the g_{ik} field. Given some arbitrary coordinate systems, each orbit of a test particle is mathematically different in each coordinate system. If we want to interpret these orbits as physically identified, different coordinate points in each system must be identified through diffeomorphisms. This identification is based only on maps between different values as functions of each coordinate system. Inside H, it is not the coordinate value (t, x, y, z) but the geometrical structure given by the metric tensor or Ricci tensor that determines the spacetime points as they are.

This view surely abandons the coordinate values as intrinsic or prior properties to identify spacetime points, but it never gets rid of properties as a manifold. The hole argument states that only with topological information, we cannot determine what orbit a particle follows inside H. For example, all orbits in Figure 3.1 are the same from topological viewpoints. They inherit common properties from the topological space and never be contradictory with each other. Their differences are seen in more concrete properties, but information about topology is common in all coordinate systems. That is to say, *topology is insufficient to specify how a body moves in spacetime.*

This means that spacetime points, which cannot be distinguished from each other only with topological properties, can be distinguished with additional features. These features should be about neighboring points included in continuous orbits a particle follows inside H, namely local relations between spacetime points. Hence, a metric is necessary to distinguish between these points, and we can specify orbits clearly in a manifold. All points have local features in the manifold, and this denies the fact that spacetime is a point manifold. However, at least discussions of the hole argument based on Earman & Norton deal with points in a manifold as if these points were devoid of metric and labeled by determinate coordinate values (t, x, y, z) in one universal coordinate system. Manifold substantivalism identifies spacetime points only with

orbits within hole H

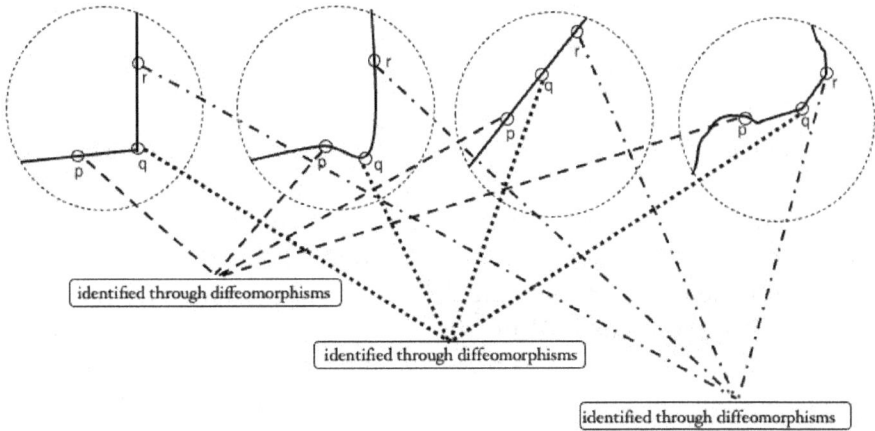

Figure 1: Apparently different orbits: This figure shows that an arbitrary coordinate system has each orbit of a test particle and seems different mathematically. But different coordinate points in each system can be related through diffeomorphisms, and if we interpret related points as the same physical spacetime points labeled p, q, r, indeterminism does not occur. This interpretation implies that all orbits are the same not only metrically but also topologically. They are different only in each coordinate value given to each point within H.

coordinate labels rather than with topological properties or positions in a manifold, leading to indeterminism. Conversely, without universal coordinate values covering a manifold, indeterminism does not occur, and the models of orbits are identified topologically. By possessing a metric, a manifold becomes so abundant that apparently different candidates with the same topological properties can also be identified as one.

Metrical information is necessary and sufficient to describe physical spacetime in this case and includes topological information. Ultimately, many orbits are identified not only by topology but also by metric. A spacetime manifold is not to be labeled by a universal coordinate system, but rather by an arbitrary one with a metric tensor defined at each point to enable us to distinguish between points.

If topology were compared to fruits, metric would be, for example, an apple, and apparently different orbits in the hole would correspond to different pictures of the same apple drawn from various directions. Even if spacetime were found to be an apple, it would not deny the fact that spacetime is a fruit. It only means that in order to draw pictures of spacetime, more concrete information of being an apple must be added.

Does metrical information always uniquely determine spacetime points in a manifold? The answer is no, and this is a weak point of structural interpretations, sophisticated substantivalism, and relationism. Even if all metrical features are given to all spacetime points in the universe, there is a case in which we cannot distinguish each of these points, namely, when some symmetries are imposed on spacetime. Wüthrich points out that if a metrical structure given to spacetime in the universe is homogeneous and isotropic, leading to the so-called Friedmann-Lemáitre-Robertson-Walker (FLRW) metric, all spacetime points will be identified (Wüthrich 2009). In other words, we cannot distinguish one spacetime point from another with metrical information depending on the symmetries imposed on spacetime. Let us now examine Wüthrich's suggestion.

The FLRW metric is derived from the cosmological principle that the universe is approximately the same everywhere and possesses translational and rotational symmetries described as this in polar coordinates:

$$ds^2 = c^2 dt^2 - a^2(t) \left[\frac{dr^2}{1 - Kr^2} + r^2(d\theta^2 + \sin^2 \theta d\phi^2) \right]. \qquad (2)$$

$a(t)$ is the scale factor of space and K is the space curvature which is constant through all spacetime points. In this metric, as Wüthrich shows, there exists a one-parameter family of spacelike hypersurfaces Σ_t, or a preferred foliation labeled by a cosmological time t (Wüthrich 2009, 1044). In addition, for any of the points $p, q \in \Sigma_t$, there exists an isometry f of the metric g_{ik} on M with $f(p) = q$. Isometries of g_{ik} on M form a group of automorphisms called the isometry group of M onto itself. This group implies a mapping between any points p and $q \in \Sigma_t$ with "all (metrical) relations fixed, that is, the ascriptions of metrical properties to "places" in the structure" invariant. By mapping, metrical properties of p can be taken over by metrical properties of q and if we follow a moderate version of OSR for spacetime points as Esfeld and Lam claim, p can be identified with q. Since spatial homogeneity enables isometries between two arbitrary points, all points can be identified in eq. 2.

> Any point on any spacelike hypersurface is thus equivalent to any other point on the same hypersurface. In particular, the symmetries imply that the spatial curvature of all the spacelike hypersurfaces Σ_t of the preferred foliation is constant. (Wüthrich 2009, 1044)

According to Leibniz's Principle of the Identity of Indiscernibles (PII), objects should be distinguished only in terms of their properties:

$$\forall F(F(a) \leftrightarrow F(b)) \rightarrow a = b. \tag{3}$$

It is important to consider how wide the range included in F is, for example, whether F is limited to physical properties or to intrinsic ones, etc. French and Redhead suggest that PII has two versions, a strong one and a weak one. In the weak version, F includes properties of spatial location, while in the strong version, F excludes properties of spatial location (French & Redhead 1988, 234)[4]. Although there are similar objects that share many properties, in classical physics, more than one rigid body cannot occupy the same space points, and they cannot be identified by the criteria of the weak PII.

Spatial location is of significant importance related to spatiotemporal properties. Structural realists consider F to consist only of automorphically invariant relational properties, not intrinsic ones. If objects in spacetime refer to spacetime points, all of them must share the same relational properties in the FLRW metric, namely 'places' in the structure. Spatial locations for spacetime points may be such invariant relational properties. Hence, it is concluded that p=q for any two arbitrary points, which is disastrous as it suggests that the (spatial) universe consists of nothing but one lonely point (Wüthrich 2009, 1040).

However, this conclusion is controversial even if we keep on being a structural realist. Aside from PII, one way to evade this conclusion is to admit primitive "numerical distinction (diversity)" such that there is more than one object with the same properties (Esfeld & Lam 2008, 33[5]). This view supports the realism of some objects in quantum entanglement and can dispense with the idea of haecceities, in other words, primitive thisness or primitive identity given to all individuals beyond the same intrinsic and extrinsic properties to ultimately distinguish each of them (Adams 1979), which opposes OSR. Intuitively, it is natural that in the FLRW metric, there are infinitely indistinguishable spacetime points rather than just one point in the universe.

Another way is to search for other properties for each object that are different from common extrinsic ones. As an example of this, Saun-

---

[4]As Wüthrich tells in his paper, French also divides PII into three versions— (i)$\forall$F ranges over all possible properties, (ii)$\forall$F ranges over all possible properties except spatiotemporal ones, and (iii)$\forall$F ranges only over intrinsic properties. (French 2006; Wüthrich 2009, 1045)

[5]In Wüthrich, numerical distinction is expressed as "numerical plurality" (Wüthrich 2009)

ders proposes "weak discernibility", which shows that an object has an irreflexive relation with itself even in a symmetric structure (Saunders 2003, 2006). He uses Max Black's example of two spheres of iron positioned in an otherwise empty universe, one mile apart in space (Black 1952), and points out that they are weakly discerned by the symmetric and irreflexive relation "one mile apart in space" (Saunders 2006, 57). If two spheres consist of the same ingredients and have the same size, shape, or color, they may surely share all intrinsic properties. In addition, both of them possess an extrinsic property of being one mile apart from the other, but they do not have a distance of one mile from themselves. So this extrinsic property is an irreflexive relation. Similarly, metrical properties are relations for a spacetime point with other points and not with itself. Spacetime points with no intrinsic properties in symmetric solutions of the field equation such as the FLRW metric are weakly discernible (Lam & Esfeld 2012, 254).

However, I would like to argue that spacetime points in a symmetric geometric structure can be distinguished by using properties from the manifold or topology. As I mentioned earlier, two spheres in Black's example, which are rigid bodies in the sense of classical physics, cannot occupy the same place and can be distinguished from each other, even if we do not appeal to weak discernibility. But if we take an extremely relationist stance, such as Leibniz, this distinction of location becomes nonsensical (Leibniz 1981). This distinction presupposes space as "a fixed background of topology $\mathbb{R}^3$" (Wüthrich 2009, 1045). The difference implies the difference in where points are put in a manifold. Although I stated that a metric field is necessary and sufficient to describe physical orbits, it does not mean that a spacetime manifold does not exist.

I think we should conclude that different points p and q in the same manifold differ only because there is a topology of spacetime, even in the FLRW metric universe. p and q are indistinguishable from a viewpoint of metric owing to isometries, but at least they are referred to by different coordinate values in the same coordinate system, which means that they are already in different places. All spacetime points in M are placed with a definite order, called topology, independently of locality. As Figure 3.1 shows, p, q, and r in each coordinate system are, at least, different points in the same hole H, independently of metrical information. Here, we want to remember Newton's quotation in 2.2.

- For the spacetime realists, the parts of space, that is to say, space points derive their character from their positions, or order.

Our universe, consisting of different infinite continuous points, has global topological features, whether a homogeneous and isotropic geometric structure is imposed on spacetime or not. Hence, two points can be distinguished from each other since they have different topological properties.

Again, I want to analogize the relation between topology and metric to the relation between fruits and apples. In this case, two objects cannot be distinguished because they are very similar apples. But originally, they were regarded as two different fruits until it was discovered that they were the same kind of fruit. This is a clear example that concrete information is not always sufficient to distinguish objects from each other.

In short, topology and metric are in a give-and-take relation to describe physical spacetime. In order to determine an orbit in a hole, topological features are not enough to know which points a particle passes through. Given metrical features, we can understand which point in a manifold corresponds to which physical spacetime point. Conversely, to tell one space point from others, which are indistinguishable from each other only with metric, topological features are needed. To describe macro spacetime in all cases, both metric and topology are necessary and sufficient.

Of course, as I mentioned in section 2, metrical properties eventually lead to topological ones, and global topology is naturally included in the structure of spacetime. However, if we regard spatiotemporal properties as features limited to locality, we cannot distinguish points that are different globally in a structure. For the structural realism of spacetime, spacetime exists as a geometric structure of a manifold/topology with metric, rather than only of metric[6].

Spacetime points may exist independently of metrical relations between them. Except for cases involving symmetric geometric structures, a spacetime point is identified only by metric, but this does not imply that a point is ontologically given by these local relations. Es-

---

[6]Sophisticated substantivalism and relationism consider spacetime to be metric rather than a manifold, and Hoefer puts forward metric field substantivalism as a refined version after he raises manifold plus metric substantivalism as an intermediate position from manifold substantivalism to metric field substantivalism (Hoefer 1996). However, he intends manifold plus metric substantivalism to refer to spacetime whose points possess primitive thisness with metrical properties. So, this position falls into hole arguments and differs from my suggestion. What I want to emphasize using the words "a manifold/topology with metric" is that not only local but also global properties for each point should be included in a structure of spacetime.

feld and Lam, following moderate structural realism, argue that "the relations and the objects that stand in the relations are on the same ontological footing and are also conceptually interdependent" (Esfeld & Lam 2008, 37). They hold that the metric tensor field defines spatiotemporal relations between spacetime points, which are necessary for the definition of the field.

Topology is *independent of locality, but it arises from metric*. Esfeld and Lam only consider metrical relations as spatiotemporal relations, and they reject topological relations as intrinsic properties. However, I believe that topological relations should also be included in spatiotemporal relations. In other words, even for structural realists, not only spacetime points but also topology or point manifold as a substructure of the spatiotemporal structure are ontologically posited without reference to local properties.

3.2   More Abstract in Micro Regions

I have shown that metric is an essential factor for defining physical spacetime, but topological features also contribute to making spacetime what it is. Roughly speaking, I adhere to a worldview in which a spacetime point or a point manifold is real independent of the locality given by the metric, but arises from the metric, and we do not need to assume primitive or intrinsic identities beyond a spatiotemporal structure. At least as far as the information about the spacetime realm is concerned, it is clear that local metrical relations are more concrete than global topological ones. However, physical theories can still be formulated without a metric, as I have discussed earlier in this section. Therefore, abstract features play an important role in defining physical spacetime as well.

A mathematical structure can approach a physical structure by adding information to itself. In the case of four-dimensional spacetime, it starts with a set of four-dimensional continuous points. Next, to form a geometric structure, these points are ordered to create each topological space covering a manifold from charts to atlases, subject to certain restrictions. At this point, global properties of spacetime are already expressed, and we can create a map of spacetime drawn on $\mathbb{R}^4$. While this structure may be abstract and insufficient to describe physical spacetime, a spacetime manifold reflects fundamental aspects of macro spacetime or the universe, which is the physical realm.

Metrical information is so concrete that it can specify physical spacetime. As noted earlier, with equation (1), there are multiple ways to move a velocity vector depending on how the Christoffel symbol $\Gamma_{\mu\nu}^{\lambda}$

is defined. Assuming metric connection, one way is fixed, but there are still infinite geodesic solutions sharing the same spatiotemporal structure. Using metrical information, we can only determine the geodesic line after choosing a specific coordinate system, which then labels each spacetime point with coordinate values. General covariance allows us to use arbitrary coordinate systems to relate a point manifold with metric to physical spacetime.

If we turn our eyes to quantum theory, the physical realm becomes more ambiguous. Some quantum gravity theories suggest that spacetime emerges from more fundamental entities (Wüthrich 2018; Huggett & Wüthrich 2013, and others). This view suggests that in micro regions, spatiotemporal features break down or no longer hold. However, how do we decide what to include in these spatiotemporal features, and what criteria should we use to claim the emergence of spacetime?

More fundamental entities vary depending on the quantum theory being used due to different methods of quantization. Quantum gravity theories are still incomplete as candidates to unify GTR and quantum mechanics. What can sophisticated substantivalism, relationism, and structural realism say about these theories? The worldviews of spacetime in a micro-region are clearly different from those in a macro-region described by GTR.

The dominant theory is super string theory, which considers units of matter to be one-dimensional strings rather than point-like particles in a fixed background of 11-dimensional spacetime. However, there is a dimensionality difference between macro spacetime and the micro background, and correspondence relations are complicated because super string theory quantizes a classical theory different from GTR, which is just a limited case only applicable under certain conditions.

The continuity of spacetime is not always maintained in other theories. Causal set theory, for example, focuses on causality within a spacetime realm but deprives spacetime of continuity. In this case, spacetime is essentially discrete and consists of many causal sets that are embedded in a manifold, rather than spacetime points. Additionally, loop quantum gravity directly quantizes the metric and arrives at a graph called a spin network, which has nodes and edges that quantify discrete volumes and discrete areas of edges corresponding to the surface of adjacency of the connected volumes, as if they describe atoms of space. In this case, spacetime is also essentially discrete. Properties that are even more fundamental than topology and metric, such as continuity, may not be necessary for a micro spacetime realm.

Local properties given by metric are not seen in micro-regions as

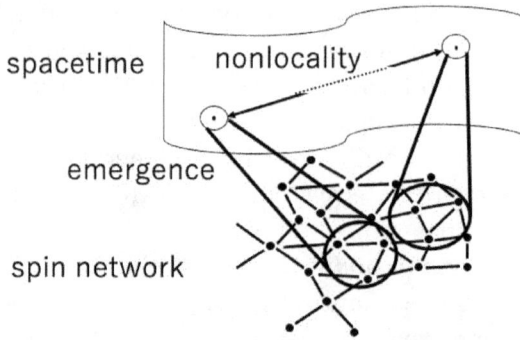

Figure 2: Emergence of spacetime from spin network: This figure shows that a spin network has locality between two nodes in a state space but exhibits non-locality as spacetime when macro spacetime emerges. I drew this figure from Fujita (2020, 7) based on the one in Wüthrich (2018, 8).

they are in a macro-region. In loop quantum gravity, as Huggett and Wüthrich show (Huggett & Wüthrich 2013, 279; Wüthrich 2018, 7-8), locality is broken in spin networks. They raise two differences between spin networks and ordinary lattices regarding locality. Spin networks are born as quantization of metric, namely spacetime, and they should express quantities about spacetime through a quantum superposition of different spin networks. Each state has its locality, and locality is not fundamental!

> Except perhaps for very special states, local beables can thus not be part of the fundamental reality, but must instead emerge in some limit—presumably the same as that in which locality emerges. How such local, i.e., topological, structures like relativistic spacetimes emerge from spin networks is at present little understood. (Huggett & Wüthrich 2013, 279)

Secondly, they point to a gap between a fundamental structure of spin network and a spatiotemporal structure about locality from a viewpoint of empirical coherence. Even in one spin network structure rather than a superposition of different spin networks, "two fundamentally adjacent nodes will not map to the same neighbourhood of the emerging spacetime...Hence the empirically relevant kind of locality cannot be had directly from the fundamental level" (Huggett & Wüthrich 2013, 279). (make reference to Figure 3.2)

Surely a spin network is a kind of state space devoid of physical features and if we read this empirical coherence with physical coherence, how does a spin network function as a physical entity?

We are now in a phase where we admit there are phenomena not presupposing a spacetime realm. Even without clear spatiotemporal features, can we talk about a physical realm? Maudlin calls this aspect "physical salience" and I think he is conscious that we must extract physical contents from a mathematical structure of a fundamental entity (Maudlin 2007). Especially for (moderate) structural realism, this question leads to ontology about whether a fundamental structure is physically real or not[7].

Many differences between a more fundamental structure and a spatiotemporal one are due to discontinuities based on correspondence relations. For example, in interpreting causal sets, as Wüthrich says, a symmetric structure in causal set theory cannot correspond to a symmetric spatiotemporal structure such as the FLRW metric in GTR, although a concept of causality is common to both theories and realms (Wüthrich 2012). In loop quantum theory, geometric features are not seen in a spin network, which is not a physical space as I mentioned above. Through quantization, many original features of a spin network form a structure, and there is no isomorphism between it and a metrical structure, even though loop quantum gravity quantizes metric.[8].

I think it is crucial to recognize the common aspects between a spatiotemporal structure and a more fundamental structure when discussing the ontology of a micro physical realm. While there may be many discontinuities between these structures, as one emerges from the other, recognizing the commonalities between them can provide a more comprehensive answer to what constitutes a physical realm beyond spacetime and what factors are the most significant in a spatiotemporal structure.

It is not true that all information about spacetime is erased in quantum physical realms. In fact, many parameters of spin networks may be isomorphic to those featuring boundary conditions imposed on the topology. Thus, a spin network can reflect a part of the information about the topology of spacetime.

To summarize, both causal sets and spin networks abstract a spatiotemporal structure. The spatiotemporal structure that arises from the metric to the manifold in classical macro spacetime is partially

---

[7]Wüthrich refers to emergence of spacetime from structural viewpoints (Wüthrich 2012, 2018), but only admits that a traditional spatiotemporal structure is no longer in circulation.

[8]I provide details based on Butterfield's formulation about emergence (Butterfield 2011) and claim that there are structural discontinuities in quantizing gravitational field while there is an isomorphism between quantized electromagnetic field and classical electromagnetic field (Fujita 2020).

transferred to the more abstract description of the quantum physical realms. In contemporary physics, including quantum gravity theories, essential aspects of spacetime consist in a pregeometric structure.

4 Conclusion: A New Way of Interpretations About Spacetime Through Macro and Micro Region

In this paper, I have focused on how a physical realm is described by physical theory in relation to the philosophy of spacetime. In this discussion, I have touched upon what scientific realism wants to claim about theoretical entities described by mathematics. Scientific realists have been studying how these abstract entities, including a spatiotemporal structure, can be said to exist. There is no doubt that we live in a concrete physical realm, namely a spacetime realm, which is very close to being "real". I have revealed the realism of this privileged realm and have claimed that the abstract features of it are also real.

In section 2, I explored the meaning of realism concerning theoretical entities by discussing the metaphysical debate between realism and nominalism. In general, causality and intervention are important factors in defining specific entities that exist in physical phenomena in our actual world, with positions in spacetime, rather than as universals in abstract models[9]. Additionally, I showed that various mathematical aspects are used to describe this privileged physical bent spacetime in GTR. In the philosophy of spacetime, spacetime is referred to by the metric field, rather than a point manifold, for empirical (physical) reasons. This has led to sophisticated substantivalism, relationism, and structural realism, which aim to solve problems such as the hole argument.

In section 3, I argue that although spacetime is metric as established by previous studies, it also has topological properties as it is a set of all points in a manifold that *arises from* the metric, independent of locality. This is important because it emphasizes that there are more abstract properties given to spacetime points than only the metric. In

---

[9]In this paper, I associated causality (an empirical property) with the spacetime realm due to the context of scientific realism. However, there is a causal realm beyond the spacetime realm in quantum theory, which is a phase space where virtual particles like photons move (Fujita 2021), even if we do not bring up quantum gravity theories as I discussed in section 3.2. Additionally, semi-particles defined in solid-state physics may be considered not real, although they have causal inertia (Gelfert 2003; Falkenburg 2007).

symmetric universes, such as the FLRW metric, spacetime points are identified not only by the metric but also by topology. In micro regions, the macro spatiotemporal structure becomes more abstract because of more fundamental entities in each quantum gravity theory. These entities reflect some features of spacetime, and a spatiotemporal structure emerges from them. Therefore, in considering quantum gravity, what is essential for being a physical realm in general is more abstract.

Through this paper, I am conscious of a realm where entities live. In a macro region, physical particular entities are in spacetime and it is presupposed in a spacetime realm that physical phenomena occur. If we call this view spatiotemporalism, to believe in micro physical realms grounding spacetime itself such as spin networks is against spatiotemporalism. However, quantum theories defend rejecting spatiotemporalism as ontology (Paul 2012) and the time has come when we should interpret physical phenomena in a broad sense.

Acknowledgment

This work was supported by JSPS KAKENHI Grant Number JP21J01573 and Research Grants of Suntory Foundation in Japan. Additionally, I want to express my gratitude to native English speakers, including my lovely wife, for checking my poor English.

References

Adams, Robert. M. (1979). Primitive Thisness and Primitive Identity. *Journal of Philosophy*, 76, 5–26.

Alexander, Horace. G, (Ed)., (1956). *The Leibniz-Clarke Correspondence: Together with Extracts From Newton's Principia and Opticks.* New York: Barnes & Noble.

Balaguer, Mark. (1998). *Platonism and Anti-Platonism in Mathematics.* Oxford: Oxford University Press.

Black, Max. (1952). The identity of indiscernibles. *Mind* 61, 153–164.

Butterfield, Jeremy. (2011). Emergence, reduction and supervenience: A varied landscape. *Foundations of Physics,* Volume 41(6), 920–959.

Cartwright, Nancy. (1983). *How the Laws of Physics Lie.* Oxford: Oxford University Press.

Chakravartty, Anjan. (1998). Semirealism. *Studies in History and Philosophy of Science*, 29(3), 391–408.

.... (2007). *A Metaphysics for Scientific Realism:Knowing the Unobservable.* Cambridge: Cambridge University Press.

Dorato, Mauro. (2000). Substantivalism, Relationism, and Structural Spacetime Realism. *Foundations of Physics*, 30, 1605–1628.

Earman, John & John Norton. (1987). What price substantivalism:The hole story. *British Journal for the Philosophy of Science*, 38, 515–525.

Einstein, Albert. (1961). Relativity and the Problem of Space. In *Relativity: The Special and the General Theory* New York: Crown.

.... (2007). Ether and the theory of relativity. In Michel Janssen, John D. Norton, Jürgen Renn, Tilman Sauer, John Stachel (Eds.), *The Genesis of General Relativity. Boston Studies in the Philosophy of Science vol 250.* Dordrecht: Springer, Volume 3, pp.613–619.

Esfeld, Michael & Vincent Lam. (2008). Moderate structural realism about space-time. *Synthese*, 160(1), 27–46.

Falkenburg, Brigitte. (2007). *Particle Metaphysics: A Critical Account of Subatomic Reality (The Frontiers Collection).* Springer.

Field, Hartry. (1980). *Science Without Numbers.* Princeton: Princeton University Press.

French, Steven & Michael Redhead. (1988). Quantum Physics and the Identity of Indiscernibles. *British Journal for the Philosophy of Science*, 39, 233–246.

French, Steven. (2006). Identity and Individuality in Quantum Theory. In Edward N. Zalta (Ed.), *The Stanford Encyclopedia of Philosophy.* Spring 2006 edition. Stanford, CA: Metaphysics Research Lab, Center for the Study of Language and Information, Stanford University.

Frigg, Roman. (2010). Models and Fiction. *Synthese*, 172(2), 251–268.

Fujita, Sho. (2020). Emergence of Spacetime from Structural Interpretations: Why is Spacetime's atom not Spacetime itself? *Journal of the Japan Association for Philosophy of Science "Kagaku Kisoron Kenkyu" in Japanese*, 48(1), 1–19.

.... (2021). Abstractness for Nominalists: Physical Worlds Outside of the Spacetime Realm (translated from Japanese) *Contemporary and Applied Philosophy*, 13: 141–173. https://doi.org/10.14989/266868

Gelfert, Axel. (2003). Manipulative success and the unreal. *International Studies in the Philosophy of Science*, 17(3), 245–263.

Giere, Ronald. (2004). How models are used to represent reality. *Philosophy of Science*, 71(5), 742–752.

Hacking, Ian. (1983). *Representing and Intervening: Introductory Topics in the Philosophy of Natural Science*. Cambridge: Cambridge University Press.

Hoefer, Carl. (1996). Absolute versus Relational Spacetime: For Better or Worse, the Debate Goes On. *British Journal for the Philosophy of Science*, 49, 451–467.

Hawking, Stephen W & George F. R. Ellis. (1975). *The Large Scale Structure of Space Time*. Cambridge: Cambridge University Press.

Huggett, Nick & Christian Wüthrich. (2013). Emergent spacetime and empirical (in)coherence. *Studies in the History and Philosophy of Modern Physics*, 44(3), 276–285.

Ketland, Jeffrey. (2011). Nominalistic Adequacy. *Proceedings of the Aristotelian Society New Series*, 111, 201–217. Oxford University Press.

Lam, Vincent & Michael Esfeld. (2012). The Structural Metaphysics of Quantum Theory and General Relativity. *Journal for General Philosophy of Science*, 43(2), 243–258.

Leibniz, Gottfried. W. (1981). *New Essays on Human Understanding*. Translated and edited by Peter Remnant and Jonathan Bennett. Cambridge University.

Newton, Isaac. (1962). *Unpublished Scientific Papers of Isaac Newton*. Translated by A. Rupert Hall and Marie Boas Hall. Cambridge: Cambridge University Press.

Maudlin, Tim. (1988). The essence of space time. *Proceedings of the Biennial Meeting of the Philosophy of Science Associationin* Vol.2, 82–91. Cambridge University Press.

.... (1990). Substances and space-time: What Aristotle would have said to Einstein. *Stud. Hist. Phil. Sci*, 21, 531–561.

.... (2007). Completeness, supervenience, and ontology. *Journal of Physics A: Math- ematical and Theoretical*, 40(12), 3151–3171.

Paul, Laurie. A. (2012). Building the world from its fundamental constituents. *Philosophical Studies*, 158(2), 221–256.

Pincock, Chris. (2007). A Role of Mathematics in the Physical Sciences. *Nous*, 41(2), 253–75.

Psillos, Stathis. (2010). Scientific Realism: Between Platonism and Nominalism. *Philosophy of Science*, 77(5), 947–958.

.... (2011). Living with the Abstract: Realism and Models. *Synthese*, 180(1), 3–17.

Rynasiewicz, Robert. (1996). Absolute Versus Relational Space-Time: An Outmoded Debate? *Journal of Philosophy*, 93(6), 279–306.

Saunders, Simon. (2003). Physics and Leibniz's Principles. In Katherine Brading and Elena Castellani (Eds.), *Symmetries in Physics: Philosophical Reflections*. Cambridge: Cambridge University Press, pp. 289–308.

————— (2006). Are Quantum Particles Objects? *Analysis*, 66, 52–63.

Slowik, E . (2004). Spacetime, Ontology, and Structural Realism. *International Studies in the Philosophy of Science*, 19 (2), 147–166.

————— (2015). The 'Space' at the Intersection of Platonism and Nominalism. *Journal for General Philosophy of Science*, 46(2), 393–408.

Stachel, John. (2002). "The Relations between Things" versus "The Things between Relations": The Deeper Meaning of the Hole Argument. In David B. Malament (Eds.), *Reading Natural Philosophy: Essays in the History and Philosophy of science and Mathematics*. Chicago and La Salle, Illinois: open court, pp. 231–266.

Suppe, Frederick. (1989). *The Semantic Conception of Theories and Scientific Realism.* Urbana: University of Illinois Press.

Teller, Paul. (1991). Substance, Relations, and Arguments about the Nature of Space-Time. *Philosophical Review,* C, 362–397.

van Fraassen, Bastiaan. C. (1980). *The Scientific Image.* Oxford: Oxford University Press.

————— (2006). Structure: Its Shadow and Substance. *British Journal for the Philosophy of Science,* 57, 275–307.

Wüthrich, Christian. (2009). Challenging the Spacetime Structuralist. *Philosophy of Science,* 76(5), 1039–1051.

————— (2012). The Structure of Causal Sets. *Journal for General Philosophy of Science* 43(2), 223–241.

————— (2018). The emergence of space and time. In Sophie Gibb, Robin Finlay Hendry and Tom Lancaster (Eds.), *Routledge Handbook of Emergence.* London: Routledge; 1 edition (March 18, 2019).

2 Imagination, Fiction and the Reality of Minkowski's Discovery of Spacetime

Gregorie Dupuis-Mc Donald

Abstract   One of the greatest achievements of modern physics is the discovery of spacetime by Hermann Minkowski. Still, talking about the "discovery" of spacetime cannot be done without further questioning its ontological status. Did Minkowski discover a real physical substrate? What is the creative role of his scientific imagination in the process of discovery? To what extent the explanatory power of spacetime supports the conclusion that it is a true description of the physical world? I consider those questions in that paper, and I claim that for Minkowski's discovery of spacetime, imagination and explanation work together in a fictionalist strategy. I explain why there is no reason to doubt the veracity of the discovery of spacetime and its physical reality.

Keywords: Minkowski spacetime, Scientific discovery, Scientific imagination, Fictions, Explanation, Realism

1   Introduction

One of the greatest achievements of modern physics is the discovery of spacetime by Hermann Minkowski. If we take for granted that spacetime offers a compelling explanation of relativity and its physical effects, then it is no surprise that its discovery should count as one of the most significant success of modern history of science (Petkov 2009). Nevertheless, I would like to suggest that talking about the "discovery" of spacetime cannot be done without further questioning its status as a scientific object. Claiming that spacetime was discovered, I suggest, implies the following two philosophical questions. Firstly, what kind

A. S. Stefanov, G. Dupuis-Mc Donald (Eds), *Spacetime Conference - 2022*. *Selected peer-reviewed papers presented at the Sixth International Conference on the Nature and Ontology of Spacetime, 12 - 15 September 2022, Albena, Bulgaria* (Minkowski Institute Press, Montreal 2023). ISBN 978-1-989970-96-6 (softcover), ISBN 978-1-989970-97-3 (ebook).

of thing is it: Is it an entity, a physical substrate (i.e. an underlying physical layer on which material entities are said to be coincident), or a mathematical model? Secondly, was spacetime discovered as a real object, or was it invented as an alternative description of the classical space-time structure of the world.

The challenge with the concept of discovery is that it involves two contradicting interpretations of the status of what was discovered. While "soft" discovery can be seen as an act or process by which a scientific construct is thought and devised, without there being any concrete evidence of the reality of the construct itself (fantasies like complex numbers, supersymmetries, twistor space), "strong" discovery can be taken as finding the existence of a physical thing (the electron, white dwarf stars, black holes) (Achinstein 2011; Penrose 2016). Accordingly, one can dispute the claim that spacetime was discovered, if discovery is understood in the strong sense. Indeed, the disagreement concerns the ontological status of spacetime (Sklar 1974). The substantival view asserts that spacetime is a physical substance that does have an independent existence of its own. In contrast, the relationist view claims that spacetime is not a real existing physical structure; only spatiotemporal entities, events and the relations between them, exist out there. The complication also stems from the fact that spacetime was not found through direct observation and experimental work, but imagined and invented by Minkowski (Corry 1997; Galison 1979; Holton 1996; Minkowski 2020b).

My plan in this contribution is to present, in section 2, the general conundrum created by the idea that scientific imagination can be taken as a source of strong discoveries about the physical world. In section 3, I outline the philosophical debate on the ontological status of spacetime, and I point out that substantivalism appears to be the appropriate position to justify the claim that Minkowski discovered spacetime. In section 4, I explain what role scientific imagination played in Minkowski's discovery of spacetime, and I show to what extent scientific imagination enables to find hidden structure in the physical world. Section 5 should provide an overview of the extent to which the spacetime structure imagined by Minkowski furnishes crucial elements of explanation of the relativistic effects we observe in the world. Finally, by showing how imagination and explanation work together in a fictionalist strategy, I intend to show in section 6 that there is no reason to doubt the veracity of the discovery of spacetime and its physical reality.

2 Imagination, reality, and scientific discovery. An overview of the problem

Spacetime can be taken as a scientific model which represents the motion of bodies and fuses the three dimensions of space with a fourth dimension of time in a single four dimensional differentiable manifold. Thus, being a model, i.e a mathematical construct used for the investigation of physical motion, it appears that spacetime was imagined and invented. After all, Minkowski himself claimed that mathematical imagination, a mixture of logical reasoning and "pure fantasies", had led him to his "radical" views on the nature of space and time (Minkowski 2020b, p. 57). Accordingly, one could think that spacetime was not discovered, in the sense used to convey that we discover things that exist but were unknown before, but was merely devised as a scientific construct, i.e. a kind of mathematical abstraction. If the latter claim is true, then philosophers have to make sense of the fact that scientists can think up hypothetical objects, properties and structures, - so-called "fictions", - that turn out to be categorized and conceived as discoveries about the real world. The philosopher has to make sense of the gap between scientific fictions and the reality of scientific discoveries.

Nevertheless, the fact that spacetime is conceived as a model does not imply that it cannot be real and have counterparts in the physical world. On the contrary, empirical evidence, the absence of conclusive refutation and the predictive successes of that model all contribute to validate the claim that spacetime is indeed the true dimension of the physical world. Accordingly, one can claim that spacetime was indeed discovered: Because spacetime is a true description of the structure of the physical world, it is the discovery of a physical thing that does exist. Yet, if we assume the latter claim to be true, it then seems odd to say that spacetime is an abstract object that has been imagined and constructed. If spacetime was discovered as being the real structure of the physical world, then it must be known to exist as a physical thing, and it cannot only be the product of Minkowski's imagination. The philosopher should make sense of the gap between the reality of scientific discoveries and the role and function of scientific imagination.

While it might seem paradoxical to assert that a scientific object was discovered by a scientist, but that the discovery came in the first place out of the scientist's imagination, I contend that the contradiction between imagination and the reality of scientific objects can be resolved if we correctly construe the role of imagination and its proper function in scientific research. Additionally, I maintain that while we can consider scientific objects (which encompass models, theories, and hypothetical entities posited through the history of science) to be the products of scientific imagination, we can conceive these objects as scientific discoveries if we justify the use of a particular scheme of inference that enables us to conclude that a hypothetical scientific object *must be true*, and thus that it can be taken as having real counterparts in the world. The scheme of inference I refer to is abduction. Abduction proposes the following scheme of reasoning: If a scientific hypothesis (defining a theory, a model, or an entity) does explain the anomalies implied by a given physical phenomenon (e.g. the abnormal result of the Fizeau experiment, the null result of the Michelson-Morley experiment), then given a genuine scientific evaluation (logical and empirical tests) of that hypothesis, we are warranted in believing that the hypothesis is true. Then, by further taking the assumption of scientific ontological realism (*true scientific propositions describe objective properties of the natural world, and we should commit to a belief in the existence of the ontology posited by scientific theories*) to be correct, we can infer that the hypothesis, by being true, also is a description of the natural world as it really is. The upshot is that because a scientific hypothetical statement is shown to be a satisfactory explanation of some physical phenomenon, we are justified in believing that the ontological content of the hypothesis consists of a discovery of something true and real about the world.

Because relativistic effects (time dilation, length contraction, Doppler effect, scattering of elementary particles) have been experimentally observed, and that none of these kinematic effects would be possible in a world which is not four-dimensional, spacetime undeniably appears to be a correct explanation of those effects. Minkowski thought that relativity and the geometric structure of spacetime accounting for it was not solely a consequence of observations, but a truly new kind of axiom or physical law, i.e a demand imposed on our mathematical equations governing physical phenomena. Thus, spacetime was indeed postulated as a first principle explaining observables (Corry 1997, p. 278). That considered, I define a scientific explanation as a definition of a correct reason for the occurrence of an observable phenomenon. Einstein wrote that we have a scientific explanation if we cease to be astonished by the occurrence of a phenomenon when we identify what will cause, or at least may possibly cause, the phenomenon. In other words of Einstein, a scientific explanation is the definition of a set of circumstances, "a real something", to which we can attribute what we observe (Einstein 2018, p. 55). That being said, if we can show that spacetime is a correct explanation of relativistic effects, then we should consider those effects as the physical manifestation of the four-dimensionality of the world[1]. Thus, the belief that spacetime is a real physical substrate along which the history in time of material particles unfolds is warranted.

I understand the word "real" to denote that a scientific object is not just a concept or an abstraction, but that it has a counterpart in the physical world which is manifested and observed through physical effects. Consequently, I don't claim that we have to commit to the ontological assumption that all scientific objects must be real, and that all scientific objects are real physical entities existing in the world. Nonetheless, if we take for granted that a hypothesis does explain physical effects that are experimentally observed by stating the scientific reasons why those effects happen, then I share the philosophical attitude according to which "the essential elements of physical theories should correspond to real objects or properties of the world" (Petkov 2012, p. 4). For the opposite attitude would be unreasonable: How could a scientific hypothesis, which is supported by empirical evidence and which explain relativistic phenomena, be false and define a structure which is unreal? Thus, because it can hardly be disputed that Minkowski spacetime does indeed provide reliable knowledge and true understanding of relativistic effects, then it is justifiable to claim that Minkowski spacetime did not just introduced a conventional description of reality; in contrast, we should indeed claim that "Minkowski consciously announced a major discovery about the world, not a discovery of a mathematical abstraction" (Petkov 2020, p. 47).

With the present work, I would like to defend the following claim: Imagination was crucial in Minkowski's formulation of spacetime, and spacetime was discovered by Minkowski as a fundamental explanation of relativistic effects. I will show why we should take the fact that spacetime does indeed explain those effects as a basis from which we can infer that spacetime must be true, and therefore real. The thesis I argue for is the following: Minkowski's discovery can be construed as an imaginative fictionalist model-based strategy. The core of my argument lies in the idea that we can reconstruct Minkowski's discovery through an abductive scheme: because spacetime explains some phenomena, we can take it as being true. If it is true about counterparts in the physical world, we can take spacetime to be real.

---

[1]For a fully-fledged argument on that point, see Petkov 2009, chapters 4 and 5.

3 The philosophical debate on the reality of spacetime

As I mentioned above, the problem is that as long as we talk about scientific discovery, - in our case the discovery of spacetime, - then we should be able to defend the claim that discovery is about something that really exists and that has physical characteristics and implications. Discovery should not be about the invention of an abstract, imaginary, scientific construct. That point is made clearly by Peter Achinstein, when he writes that what discovery is about "is discovering some physical thing or type of thing (such as the electron, the Pacific Ocean), rather than discovering some abstract object (such as a proof)" (Achinstein 2001, p. 268). Hence, talking about scientific discoveries should imply that we have strong evidence of the reality of the object discovered. I will adopt Achinstein's position: "Discovering something requires the existence of what is discovered. You cannot discover what doesn't exist (...)" (Achinstein 2001, p. 268).

That being said, the philosophical debate on the reality of spacetime revolves around substantivalism, relationism, and conventionalism. The substantival view is as clear as crystal about its ontological commitment to the existence of spacetime as a real physical entity. Indeed, that view proposes to take spacetime as a physical substance with an independent reality (Sklar 1974, p. 161). The word "substance" suggests that spacetime carries and supports space and time relations between objects. Thus, spacetime can be taken as a structure that exists and that has specified features independently of the existence of any ordinary material objects (Sklar 1974, p. 161). Our universe, even without matter, would be a universe with a four-dimensional structure, where time flows, and locations are ordered in space. The crucial point is that spacetime not only has an independent existence from matter, but also that that structure is responsible for the relativistic effects we observe in the motion of matter. What I would like to show is that the substantival view is coherent with Minkowski's own ontological conception of spacetime.

Sklar writes that the substantivalist position is a natural position to hold since it can be read off from the scientific discourse on relativity. Sklar explains that once the spacetime structure of the world is defined, it is correct to assume that objects move, and that time flows, "in" spacetime (Sklar 1974, p. 164). It is the fact that events are given "in spacetime", and not with respect to an absolute space and an absolute time, that allows to trace the motion of physical object as building blocks entirely given in an independent four-dimensional structure, and to realize the absence of absolute uniform motion. Thus, the observation of relative motions can be seen as being dependent on the existence of a fundamental spacetime structure of the world. Yet, a counterposition to the substantival view, called relationism, asserts that the existence of spacetime is merely an illusion. According to that view, what there is are material entities, and material events, and what is fundamental are the spatiotemporal relations between objects. The idea of a spacetime structure existing independently is a confusion, since spatiotemporal relations exist only insofar as there are concrete events that happen in the world. Events do not happen in spacetime, but rather spatiotemporal relations are possible only because material points can be taken as happening in relation to each other.

I contend that if we take the relationist view of spacetime, then it is difficult to support the idea that spacetime was discovered as a real structure carrying events in the world. Indeed, the relationist view seems to suggest that spacetime is merely a convention used to describe spatiotemporal events, and that other alternative frameworks would be equally good to furnish the conceptual means for the description of spatiotemporal relations between objects. In fact, the conventionalist view asserts that when there is a competition between equally good theories and that we cannot discriminate among those theories on the basis of their observational consequences, we are free to decide, by convention or decision, which theory we want to use in order to save the phenomenon (Sklar 1974, p. 121). Usually, we choose the theory that is the most convenient in terms of simplicity and utility, and we accept that theory as true by convention. It should be noted that the conventionalist position was endorsed by Poincaré with regard to spacetime. Indeed, Poincaré showed that the standard geometric operations needed for relativity theory could be performed in an alternative four-dimensional framework. Poincaré showed that the Lorentz transformations could be carried out like a rotation as if ordinary space and time were combined in a four-dimensional formalism. Poincaré knew that "it would be possible to translate our physics into the language of geometry of four dimensions" (Poincaré 2017, p. 427). Nevertheless, Poincaré thought that the four-dimensional structure was just a matter of mathematical convention, and not a matter of a real physical substrate. He also believed that it was an inconvenient convention. Indeed, he writes: " To attempt that translation (to translate our physics into the language of geometry of four dimensions) would be to take great pains for little profit. (...) It seems that the translation would always be less simple than the text, and that it would always have the air of a translation, that the language of three dimensions seems the better fitted to our description of the world" (Poincaré 2017, p. 427). We can see that Poincaré did not endorse a substantivalist position on spacetime. As Petkov emphasizes, Poincaré failed to comprehend the profound physical meaning of the four-dimensional framework, and by neglecting the thought that spacetime can be seen as real physical dimension of the world, certainly was prevented in being attributed the discovery of spacetime (Petkov 2020, p. 22).

4 The role of explanation for Minkowski's discovery

All that being considered, the question that remains to be answered is the following: What is the explanatory power of Minkowski's spacetime which supports the conclusion that spacetime is a true description? If it turns out that the explanatory power of spacetime is so obvious that it makes it unreasonable to deny its truth, then the argument suggests that spacetime is not just an imagined fiction, but a true description of reality. Consequently, the explanation provided by spacetime shall be taken as a real discovery about the physical world.

As remarked above, Minkowski imagined a model in which events are represented in space and in time through a particular value of a coordinate quadruple x, y, z, t. An event is thus represented by a point of space at a particular time, which is called a world-point. Minkowski further suggested that we could imagine all world-points of a singular object as tracing the trajectory of an object in spacetime through a world-line. For Minkowski, the set of all possible systems of values of these world-points constitutes the world, and world-lines are trajectories in that world. Accordingly, the world is made of as many infinite spaces as there are world-points, where time is individually given as a fourth dimension for every-world-point, defining the different moments of the existence of an object. We can see that spacetime provides a natural explanation of the requirement imposed by relativity that all laws of nature be the same in every reference frame. Indeed, since a world-point can be associated with an inertial observer in her or his reference frame, what the observer perceives is singular to that world-point, i.e. can be described in terms of her or his own space and time in which she or he is at rest. In other words, every inertial observer describes what she or he perceives according to the same exact physical laws, in terms of the space and time in her or his reference frame. For the same reason, Minkowski spacetime explains why an observer at rest in an inertial reference frame cannot detect absolute motion. The fact that the observer is at rest at a singular world-point in spacetime makes it impossible for her or him to determine his motion *in* space, because there is not one privileged space with respect to which the observer can compare her or his motion, but an infinite set of possible values for other coordinate quadruples.

Moreover, it appears that while the observer should confirm that the time it takes for light to travel is always the same in her or his spacetime reference frame, it will be noted that since two observers have different times corresponding to their respective frames, then two observers can disagree on the time it takes for light to travel as observed in the other's reference frame. If we consider the world-lines of objects in spacetime, the fact that time is not absolute also finds a natural explanation: it is a consequence of the structure of spacetime. Indeed, since all moments of the time of every particle is entirely given in spacetime through the trajectory of its world-line, the world-lines of two different particles are distinct, in such a way that observers at rest can choose their time axes along a given world-line. Consequently, we see that the time of an observer at rest with respect to another world-line is relative to the other world-line. The comparison of the observers' time cannot be made with respect to an absolute time, but to the time given in their respective frames of reference. In spacetime, the shape of world-lines, i.e. their inclination of curvature, thus indicates the extent to which the motion of inertial frames with respect to another influences the perception of the events that are happening in every reference frame. The inclination of a world-line with respect to another involves a distance accounts for the dilation of time; also, the curvature of a world-line with respect to another makes it clear that the distance between the reference frames changes with time because of acceleration.

These two explanations show that we can attribute the physical meaning of the relativity principle and some of its effects to the structure of spacetime. Accordingly, it appears that spacetime is not just a description of relativistic effects, but the element that tells why these relativistic effects appear and are observable. Thus, spacetime enables to understand the observable effects of relativity. As the two examples above show, spacetime explains the two postulates of relativity, i.e. that natural phenomena run their course according to the same general laws in all reference frames, because the real times in all these frames can be treated equally, and that the velocity of light is constant. Yet, while spacetime tells that the proper times of the observers will not differ if they cover the same distance in inertial motion, it explains why the observers can disagree about the time in another frame, as measured from the other. Because spacetime tells that time is dependent on space, we see that the fact that the other world-line is inclined because it is in another frame which is not parallel to the other observer, we conclude that the distance between the spatial components of the frames as projected in the other frame implies that time will be observed as dilated. That is attributed to spacetime since it tells us that there is not just one absolute time, but that there are many times.

5 The role of imagination in the discovery of spacetime

I gave an overview of the debate about the reality of spacetime, and I suggest that a reconstruction of Minkowski's discovery of spacetime is coherent with a substantivalist view. Yet, as I explained in the first section of the paper, the question concerning the role of imagination in the formulation of spacetime, and the fact that one could argue that spacetime was invented as a mathematical construct, needs to be addressed. The problem, to repeat, is the following: If spacetime was discovered as being the real structure of the physical world, then it must be known to exist as a physical thing, and it cannot be the product of Minkowski's imagination. We need to bridge the gap between the reality of scientific discoveries and the role and function of scientific imagination.

In any case, Minkowski himself claimed that mathematical imagination was the ability which had led him to arrive at his conception of spacetime. He describes mathematical imagination as the habit of mathematicians to see a given problem through different points of view. In other words, mathematicians have the capacity of analyzing if different theoretical structures are equivalent, or if they differ in their mathematical and physical consequences. Minkowski saw how different groups of geometrical transformations could be approximated mathematically while they had disparate physical implications. Indeed, Minkowski showed that the Galilean group of geometric transformations appropriate to Newtonian mechanics, where the x-axis is left fixed and the t-axis is completely free such that all frames agree on simultaneity, was just a limiting case contained in the more general group of Lorentz-transformations. Minkowski considered the latter group by introducing an additional parameter c of the finite speed of light in the graphical representation of the rotations of space and time around the origins of coordinates. Minkowski demonstrated that the latter group of transformations was more intelligible and theoretically more satisfying considering the new developments in theoretical physics. Yet, he did so by stressing that privileging that group of transformations implied a belief that spatio-temporal phenomena manifest themselves in terms of a four-dimensional world. Minkowski called the latter implication the "world-postulate". He stressed that the world-postulate was not only more convenient for a symmetrical treatment of space and time coordinates, but also to show in a novel fashion how the true form of the laws of physics appear. Accordingly, Minkowski could claim that the whole world presents itself through that structure (Minkowski 2020b, p. 112).

The strength of Minkowski's imagination lies in the fact that the physical consequences of the Lorentz-transformations could be understood through visualization. Minkowski found how the algebraic relations of the Lorentz-transformations could be visualized through a geometric representation. The geometric visualization of the physical consequences of the algebraic features of the Lorentz-transformations, illustrated by the hyperboloid diagram on the basis of which Minkowski could further introduce the concepts of world-lines and light-cone, facilitated the intuition and understanding of the concept of spacetime and its implications. Still today, the visualization of specific geometrical relations between lines in four-dimensional spacetime diagrams are crucial in understanding why relativistic effects like time dilation, length contraction and the relativity of simultaneity are consequences of the geometric structure of spacetime. Consequently, imagination is not only a creative ability; its visual component has an explanatory power. Because what is imagined can also be communicated through visualization, the imagination of a scientist can be understood while being shown to others, and the others can see and become aware of special relations that could not be seen before. Hence, Minkowski's imagination can be seen as an essential factor in the reception of the discovery of spacetime.

Minkowski claims that mathematicians have the capacity to open new territories of investigation "within their pure fantasies" (Minkowski 2020a, p. 39). Mathematical imagination is a realm within which new facets of the physical world can be discovered. If it can be shown that mathematics interprets and corresponds to physical phenomena, we witness the power of the application of mathematical imagination to the physical world. In other words, "the visualization of nature's laws through geometry enters as the primary motivation for the creation of a new physical and metaphysical outlook" (Galison 1979, p. 117). For Minkowski, that the world in space and time is a four dimensional, non-Euclidean manifold, should be seen as "almost the greatest triumph that the application of mathematics has brought about as of today" (note relativity principle). Furthermore, mathematical imagination is not just a description of the physical world; the mathematician's fantasies "contain the most complete real existence" (Minkowski 2020a, p. 39).

That being considered, the "thematic" aspect of Minkowski's imagination should be stressed. Thematic imagination refers to the core presuppositions and beliefs a scientist holds (Holton 1996, p. 201). Those presuppositions and beliefs are linked to a scientist's imagination because they circumscribe her or his thoughts on what exists, and what is fundamental. Hence, they are individual attitudes towards a specific scientific content and they shape individual beliefs concerning the status of specific objects. They should consequently be distinguished from paradigms, or research programmes, that are not linked to the faculty of imagining, but to theoretical and methodological guidelines of a whole community. Thematic imagination includes a scientist's core conceptual and ontological choices, and these are found in the way a scientist ranks entities, properties and relations according to their status in a theory.

Accordingly, "themata" in science are the individual preferences and commitments that scientists adopt that constrain and motivate research (Holton 1975). For example, in the context of the philosophy of spacetime, "presentism", the view that holds that what the world is the present as defined as everything that exists simultaneously at the present moment, as opposed to four-dimensionalism, which in contrast holds that the world is timelessly existing, time being already given, where the past and the future are already mapped out, constitute opposite themata. They are opposite, because if we consider the same thought experiments, but through the two different themata, we imagine scenarios that are totally different. Another example of the function of thematic imagination in providing fundamental thinking categories can be seen in the opposition between "relationism" and "substantivalism" in the philosophy of spacetime. Relationism asserts that while matter is what exists in the world, spacetime is an abstraction realized by metrical relations between material bodies. Substantivalism, in contrast, claims that spacetime is a "substance", and that that substance does exist independently of matter. The commitment to either of these themata indicates how profoundly they can influence how one can imagine the structure of spacetime and how they determine the description of properties of matter.

Concerning Minkowski's thematic imagination of spacetime, the crucial point is that he believed that spacetime is the true dimension of the world, and that physical events do happen in a four-dimensional world. For Minkowski, what we perceive are not objects in a unique three-dimensional space, but past images of objects that existed before at various distances in different three-dimensional spaces belonging to different moments of time. The objects we perceive are not evolving in a unique absolute space, but in multiple sections of three-dimensional spaces that are fragmented by time. Thus, the existence of spatially extended objects in the world must be imagined as a set of events containing different three-dimensional entities at all given moments in time of their histories. Consequently, we can visualize an object only by imagining its existence as being resolved through all its space points at different moments of time in spacetime. In other words, the identity of an object can be recognized by the identity of a substance in all time elements through the changes of its worldpoints in space. For Minkowski, we obtain "an image, so to say, of the eternal course of life of the substantial point, a curve in the world, a worldline" (Minkowski 2020, p. 112). Accordingly, the laws of physics, and the natural world itself, can be seen as resolved through these worldlines in spacetime.

6 How Minkowski's discovery can be taken as the outcome of a fictionalist strategy

As it was explained in the first section, it is difficult to assume that imagination alone can lead a scientist to a scientific discovery. Indeed, we have to admit that for a scientific construct (theory, model, hypothetical entity) to be counted as a scientific discovery, it should provide a true description of some aspects of the world, such that its confirmation should reveal that it has physical counterparts that are real features of the natural world. From the considerations of Minkowski's imagination presented above, we can claim that Minkowski believed that spacetime was indeed a true description of the real dimension of the world, and that Minkowski's imagination makes us believe that it is the case. Nonetheless, that does not suffice to claim that Minkowski made a discovery about the physical world. It could be the case that Minkowski found an innovatory way to describe the abstract dimension of space and time in accordance to physical knowledge, but that spacetime is only a conventional way to frame physical claims in terms of a four-dimensional geometry.

What I would like to show in the following section is that imagination plays a crucial role in inventing and presenting scientific models, and that imagination has a function for scientific discovery if we conceive it as providing models that explain what we observe of physical reality. I would like to claim that if an imagined model does explain some aspect of the natural world, then it should be taken as a true model. By being true, it is rational to believe that what the model describes has real counterparts in the physical world, as long as evidence supports that belief. To think the contrary would be irrational, and contradicting evidence when there is some. I will develop that claim by presenting how scientific imagination fits into a fictionalist model-based strategy.

The problem with scientific imagination considered in itself is that it does not provide any clue as to the truth of the imagined object. When we imagine and think about an entity, a model or a theory, nothing in the act of imagining it provides us with reasons to believe that it is true of the world. In other words, even if a scientist has a priority over a certain object by being the first to have imagined it, that is not enough to classify that object as a scientific discovery, and to claim that the scientist is the discoverer. In fact, when leading scientists engage with imagination in order to visualize a scientific object and think about its implications, what is in play is a type of attitude toward a specific theoretical scenario. In other words, imagining the existence of physical bodies involved in relativistic effects in a three-dimensional space, or in a four-dimensional spacetime, is thinking about these objects *as if* the world could have different geometrical dimensions. Yet, even if one scenario happens to be theoretically more satisfying than another, nothing prompts us to believe that what is imagined is true. Therefore, we can imagine as many scenarios as we want without having to believe that one scenario constitutes a discovery about the world (Levy and Godfrey-Smith 2020, p. 5). Whoever is engaged in imagining that scenario is not logically required to believe that the scenario is true about the world, nor that the entities involved in the scenario are real and have counterparts in the natural world. Consequently, imagining the scenario in question with all its consequences cannot be seen as providing reasons for thinking that what was imagined constitutes a scientific discovery. As long as we take a realist standpoint on scientific discoveries, then we cannot take the products of imagination, things that are fictional, as true descriptions of the real world without further reasons to make us believe that what they describe is indeed the case.

As Salis and Frigg point out (Salis and Frigg 2020), while scientific imagination can be distinguished according to its content, which can be "propositional" or "objectual", what characterizes imagination is the epistemological attitude we have towards the content imagined. It must be noted that we can imagine a proposition, for example "an event happens in spacetime", and we can imagine an object, for example "a curved line in a light cone", but both varieties of imagination can be taken as being propositional because imagining an object can be seen as entertaining, or thinking about, the concept of the object, without forming a mental image of the object.

According to the authors, there is a minimal core of propositional imagination (Salis and Frigg 2020, p. 30). It contains three features: Freedom, mirroring, and quarantining. Freedom is the feature that accounts for the fact that we are free to imagine whatever we want. As far as rationality is concerned, we are not free to believe whatever we want. To believe a proposition about a scientific object is to believe that proposition to be true, and the state of affairs that makes that proposition true does not depend on us. Yet, we are free to imagine that proposition, whether or not that proposition is true. We can think about it without commitment to its truth. Nonetheless, mirroring implies that if we imagine a given scenario and its consequences, then we should be committed to the inference that brings us to these consequences. Put differently, the inferences brought by an imagined scenario cannot be ignored whether or not the scenario is real. Finally, quarantining suggests that imagining a proposition does not entail believing that proposition. If we imagine a certain scenario, then that scenario is valid as long as it is quarantined by a certain context. In other words, because the scenarios are imagined, they are set apart in a fictional context, and consequently they do not engage to believe the scenario to be the case. Thus, for Salis and Frigg, the minimal core determining the necessary and sufficient conditions for something to be an instance of scientific imagination suggests that imagination does not imply the commitment to the truth or the existence of the content imagined. The upshot of the analysis of the minimal core is that scientific imagination does not imply commitment and belief of the truth of a given scientific object. Thus, claiming that scientific discovery can be the outcome of imagination is one step removed from describing what imagination is.

That being said, if imagination, by itself, is not a ground for taking something to be true, and to believe that it is a true description of the world, then we need to define its function in facilitating scientific discovery. If the imagination alone cannot provide us with the proofs of a scientific discovery, then its role in pointing to possibilities for new discoveries should be defined. In fact, if we are looking for the confirmation that an imagined scientific object should be considered as a scientific discovery, then we should consider to what degree that object tells us anything true about the actual world. Indeed, it might be that imagining scientific objects is one way to consider what scenarios provide a better understanding of the world. If imagining a scenario helps to explain a given phenomenon, then we shall assume that the imagined object leads to the discovery of what is involved in the explanation.

The preceding discussion brings to the fore the idea that scientific objects that are thought and presented through the imagination can be taken as fictions. We can see them as fictions, because when a scientist develop a model or a mathematical structure from imagination, that model or mathematical structure can be taken as purely hypothetical. As we just saw, the minimal core of propositional imagination puts the commitment to the truth of imagined objects outside of the scope of imagination. However, independently of the way the scientific object is described, be it through a specific formalism, concepts or diagrams, it should be taken for granted that the object is indeed the description of an actual real-world system, yet not necessarily a true description.

The model-based science strategy has been put forward by Peter Godfrey-Smith and Roman Frigg (Godfrey-Smith 2005; 2009; Frigg 2009) to make sense of the fact that many scientists, to tackle scientific problems, come up with models that are "imagined physical systems" (Frigg 2009, p. 253). The scientific object is seen as an "imagined concrete thing" (Godfrey-Smith 2005, p. 734). In other words, before it is demonstrated that the model has explanatory power, and that the model has some empirical support, then the model remains imaginary or hypothetical, but it would "be concrete if it was real" (Godfrey-Smith 2005, p. 735). Hence, that strategy tries to account for the fact that before coming to genuine and actual scientific discoveries about the physical world, scientists usually present a hypothetical system as object of study and then demonstrate how that system explains some particular part of the world. The upshot is that when scientists talk about clocks, measuring-rods and rotating discs, these entities are fictional because they are imagined objects in a hypothetical situation. Nevertheless, the entities involved in the competing models can be taken as real physical things. Also, what a model explains about the world can be believed to occur in reality. If the model starts as a "fiction, a creature of the imagination", once it is shown that it does explain the target phenomena it is supposed to describe, then the model, if it existed, could be seen as a "concrete, physical thing" (Godfrey-Smith 2009, p. 101). As such, it can be concluded that the model provides understanding of the natural world. With respect to scientific discovery, what confirms that an imagined model can be considered as a scientific discovery is the fact that the model provides an explanation of a problematic aspect of the world. In other words, as long as a model provides understanding of a specific aspect of the natural world, what has been discovered is not fictional, but a novel explanation of a given phenomena through an innovatory model. The function of imagination is thus the medium with which explanations of a given phenomenon are created and represented.

If a model does explain a certain phenomenon, then the model can be taken as true. A scientific explanation is satisfying if it is grounded on empirical evidence, if it defines the factors that define why a phenomenon is observed, and it should have a certain predictive power. Those criteria suffice to rule out the objection that there are always multiple things that could explain a given scientific problem. The theoretical and empirical context together with the fact that revolutionary ideas and paradigms are not easily accepted in scientific communities always reduces the set of possible explanations to a very few competing ones. Consequently, it must be stressed that it would be unreasonable to admit that a given model is an explanation of a phenomenon, but to deny that the entities and relations the model posits do not have counterparts in the physical world. To be sure, empirical evidence has always the last word in confirming if a model is indeed a correct representation of the natural world. But if there is no empirical evidence *against* a given model that satisfyingly explain a given phenomenon, then it is unreasonable to admit that the model is successful and resolve open questions, but to doubt that the model reflects real properties of the world. In contrast, it seems that the correct attitude to have toward that model is to conclude that the model is true, and by being true, we should conclude that what it describes has analogs in the natural world. For the opposite attitude is unsound. If one admits that a model is a correct explanation of a certain phenomenon, then one is bounded to acknowledge the truth of the model. Without any empirical evidence to the contrary, it would be unreasonable to claim that the model does not represent real properties and objects of the physical world. Hence, if one doubts that a model does explain a given phenomenon, but claims that the model is not adequate, then one should be able to point out what theoretical entities are not essential to the model, or do not in fact exist, and why another competing model can discard the latter model because it describes and explains the phenomenon more adequately. As such, the burden of proof is on the one who is in need of refuting a given theory, who needs to find evidence refuting the given model, a not on the one who provides a satisfactory explanation of a scientific problem.

7 Conclusion

The larger problem studied in that paper is to justify the claim that Minkowski discovered spacetime. The challenge stems from the realist position, and lies in finding an argument that supports the view that Minkowski discovered a real physical substrate that exists out there in nature. As I emphasized, the difficulty is also rooted in the fact that scientific imagination was crucial for the definition and development of the fiction of spacetime. Thus, my objective was to bridge the gap between a realist and a fictionalist approach to scientific discovery. I attempted to answer two questions: Firstly, what kind of thing is spacetime: Is it an entity, a physical substrate (i.e. an underlying physical layer on which material objects are said to be coincident), or a mathematical model? Secondly, what grounds the claim that spacetime was discovered and is a real object, and not, in contrast, that it was invented, or suggested as an alternative fictional description of the classical space-time structure of the world. I suggested that the answer to the first question requires the substantivalist view of spacetime. According to it, spacetime is a physical substance with an independent reality, such that it carries and supports space and time relations between objects. I showed how that view is coherent with a realist conception of scientific discoveries. Concerning the second question, my proposal is that the explanatory power of spacetime to explain why relativist effects are observable in the world supports the inference that spacetime must be a true description of reality. Since it is assumed by Minkowski, and by the substantivalist view, that spacetime is a real substrate, and not only an abstract mathematical model, I claimed that by being true, spacetime must be real. I focused on the fictionalist strategy to underscore the fact that while imagination is the realm of creation of abstract scientific models, the explanatory power of those models together with the empirical evidence of its truth allows one to infer that the fictional constructs implied in the model description can be taken as real natural objects of nature.

References

Achinstein, P. (2001) *The Book of Evidence*, Oxford University Press, 304 p.

Corry, L. (1997) "Hermann Minkowski and the postulate of relativity", *Archive for History of Exact Sciences*, Vol. 51, pp. 273-314.

Einstein, A. (2018) *Relativity*, Minkowski Institute Press, 173 p.

Frigg, R. (2009) "Models and fiction", *Synthese*, Vol. 172, pp. 251-268.

Galison, P-L. (1979) "Minkowski's space-time: From visual thinking to the absolute world", *Historical Studies in the Physical Sciences*, Vol. 10, pp. 85-121.

Godfrey-Smith, P. (2005) "The strategy of model-based science", *Biology and Philosophy*, Vol. 21, pp. 725-740.

Godfrey-Smith, P. (2009) "Models and fictions in science", *Philosophical Studies*, Vol. 143, p. 101-116.

Godfrey-Smith, P. and A. Levy (eds.)(2019) *The Scientific Imagination*, Oxford University Press, 360 p.

Holton, G. (1996) "On the art of scientific imagination", *Daedalus*, Vol. 125, No. 2, pp. 183-208.

Holton, G. (1975) "On the role of themata in scientific thought", *Science 25*, Vol. 188, Issue 4186, pp. 328-334.

Levy, A. and P. Godfrey-Smith (ed.) (2020) *The Scientific Imagination*, Oxford University Press, 337 p.

Levy, A. and P. Godfrey-Smith (2020) "Introduction" in *The Scientific Imagination*, Oxford University Press, pp. 1-16.

Minkowski, H. (2020a) "The relativity principle", in Petkov, V. (ed) *Spacetime. Minkowski's Papers on Spacetime Physics*, Minkowski Institute Press, 212 p.

Minkowski, H. (2020b) "Space and Time", in Petkov, V. (ed) *Space-time. Minkowski's Papers on Spacetime Physics*, Minkowski Institute Press, 212 p.

Minkowski, H. (2020c) "A derivation of the fundamental equations for the electromagnetic processes in moving bodies from the standpoint of the theory of electrons" in Petkov, V. (ed), *Spacetime. Minkowski's Papers on Spacetime Physics*, Minkowski Institute Press, 212 p.

Penrose, R. (2016) *Fashion, Faith, and Fantasy*, Princeton University press, 491 p.

Petkov, V. (2009) *Relativity and the Nature of Spacetime*, Springer international, 329 p.

Petkov, V. (2012) *Inertia and Gravitation*, Minkowski Institute Press, 151 p.

Petkov, V. (ed) (2020) *Spacetime. Minkowski's Papers on Spacetime Physics*, Minkowski Institute Press, 212 p.

Pyenson, L (1979) "Physics in the shadow of mathematics", *Archive for History of Exact Sciences*, Vol. 21, pp. 55-89.

Poincaré, H. (2017) *Science and Method*, Cosimo Classics, 292 p.

Salis, F. and Frigg, R. (2020) "Capturing the scientific imagination", in Godfrey-Smith, P. and A. Levy (eds.)(2019) *The Scientific Imagination*, pp. 17-50.

Sklar, L. (1974) *Space, Time, and Spacetime*, University of California Press, 400 p.

3 EINSTEIN AND WITTGENSTEIN'S LADDER

ELTON MARQUES

Abstract   The question on whether simultaneity is absolute or relative, contrary to what one might think, did not close with the success of special relativity. Different interpretations of this theory – or different versions of it – still arouse interest in many philosophers of science, some scientists and a vast number of metaphysicians. Usually, those who challenge its more widely accepted version point out the empiricist and verificationist commitments at the heart of this theory. Since Einstein himself acknowledges empiricism, identifying flaws in the underlying epistemology of this theory would be a way of challenging the beliefs shared by classical versions of relativity, such as the relativity of simultaneity. However, is showing the weak empiricist foundations of this theory sufficient for challenging well-established results? In this article, I set out to defend that, despite Einstein's notorious endorsement of empiricism, one is not required to understand the result known as the 'relativity of simultaneity' as dependent on any form of verificationism. Therefore, I attempt to provide a defence of relative simultaneity while accounting for its purported advantages, as well as the difficulties identified by its critics. After all, Einstein's empiricism, especially his verificationism, might simply be a kind of 'Wittgenstein's ladder' - less fundamental than one might think.

Keywords: relativity; simultaneity; verificationism; empiricism; time

1 Introduction

Einstein's achievements have been connected to several philosophical questions. Among them, the question of simultaneity – of whether or not it is relative – is particularly important. It is a significant question by virtue of the fact that how we answer this question has important

A. S. Stefanov, G. Dupuis-Mc Donald (Eds), *Spacetime Conference - 2022*. *Selected peer-reviewed papers presented at the Sixth International Conference on the Nature and Ontology of Spacetime, 12 - 15 September 2022, Albena, Bulgaria* (Minkowski Institute Press, Montreal 2023). ISBN 978-1-989970-96-6 (softcover), ISBN 978-1-989970-97-3 (ebook).

consequences for a number of philosophical debates, such as the issue of eternalism versus presentism, tensed versus tenseless theories, fourdimensionalism versus threedimensionalism, perdurantism versus endurantism, etc.

These debates gained renewed momentum with Einstein's 1905 paper, 'On the Electrodynamics of Moving Bodies'. Since then, the relativity of simultaneity – a consequence of the postulates of Special Relativity – has been endorsed and rejected, as has been the philosophical theses which depends on its theoretical acceptance.

The rejection of the relativity of simultaneity (henceforth, *RoS*), although surprising, is quite common, especially within specifically philosophical circles. Authors such as Craig (2002, 2008), Quentin Smith (2008), Markosian (2004), Čapek (1965), Valentini (2008), Tooley (2008), Callender (2008), Builder (1971), Ives (1979), Prokhovnik (1985, 1987, 1988), Balashov (2000) and many others, from several philosophical traditions, reject *RoS* and endorse different interpretations of the theory.[1] Some of these authors suggest that there might be a privileged (or absolute) referential, relative to which the speed of light is always c. As such, the speed of light will always be c only in those referentials that are at rest relative to the absolute referential.[2] In referentials that move relative to the privileged one, the light signal will undergo a delay or acceleration depending on the direction of the movement. Notably, the element that plays the role of the absolute referential may change from author to author, such as luminiferous ether – the compensatory response of ether; Newtonian absolute space; divine perspective – capable of distinguishing absolute from relative time; Robertson-Walker's metric dubbed cosmological fluid; microwave cosmic radiation – remnant of the beginning of the universe; vacuum of

---

[1]Some of these authors seem to reject Einstein's relativity by adopting a second theory. However, are there two or more rival theories of relativity or merely different interpretations of the same theory? Trying to answer this question would force us into an important digression, to the extent that we would have to think about the conditions under which 'theory' versus 'theory interpretation' are defined. Here, we will simply stipulate, without prejudice against any known information on special relativity, that these are different but valid and interesting interpretations of the same theory, in the same sense as there are many known interpretations of quantum mechanics, such as the Copenhagen interpretation, Everett's multiverses, etc.

[2]Following this reasoning, any referential at rest relative to absolute space would also be a privileged referential. Thus, when ether is supposedly at rest relative to absolute space, it would also be a referential relative to which we could theoretically detect absolute movement. Meanwhile, attempts at detection have failed systematically, as was the case with the Michelson-Morley experiments. (cf. Michelson and Morley 1887).

quantum electrodynamics – as proposed by Dirac (Dirac 1951, 906-7), etc. Authors who endorse some version of these interpretations (which I collectively refer to under the label of 'Lorentzian')[3] are convinced that proposals that preserve absolute simultaneity have at least the following virtues: a) empirical equivalence, meaning there is at least as much evidence for Lorentzian interpretations as there are for the interpretations accepting *RoS*; b) a metaphysical advantage, indicating that the Lorentzian version is superior from the point of view of its associated metaphysics, since it preserves common sense without introducing any revolution or break with Newtonianism; and c) a better associated epistemology, suggesting that the versions that hold *RoS* would suffer from severe faults due to their underlying verificationism.

In the following sections, we shall discuss the supposed advantages accruing from Lorentzian interpretations, with special emphasis on the alleged dependency of *RoS* on empiricist verificationism inherited from Mach (1960) and others. Subsequently, a defence of the interpretations accepting the *RoS* is motivated.

2 Relativity of simultaneity and divergent voices

RoS appears as a consequence of the unrestricted acceptance of the following postulates, which conflict with Galilean transformations and the velocity addition law: a) the laws of nature are the same in all inertial reference frames, and b) the speed of light in free space has the same value in all inertial frames of reference in all directions. Meanwhile, given Galilean transformations and the velocity addition law, either the laws governing phenomena are different for different inertial observations, or the speed of light should vary with the speed variation of the light emitting source or between different systems of coordinates.

---

[3]There are three main families of interpretation explored in the literature – two threedimensionalists and one fourdimensionalist – named by Craig as Einsteinian, Lorentzian and Minkowskian, respectively (Craig 2008, 12-16). The difference between these families mainly concerns their relativistic effects, which Einsteinians consider as real. In Minkowskian interpretations, *RoS* and other relativistic effects are merely perspectival, as illustrated when geometry is attributed to space-time. No results in these interpretations are compatible with Newtonianism, which effectively cannot be salvaged. In contrast, Lorentzian interpretations search for a compensatory effect for *RoS*, such as time dilation and length contraction, and are compatible with the Newtonian distinction between absolute time, which is independent of our measurements, and relative time (cf. Newton 1990). New interesting interpretations have also been proposed, such as Kit Fine's (2005) fragmentalism. However, it is not our intention to be exhaustive on the possible interpretations of the theory's formalism, some of which differ from the schemes presented above.

A third option, which is not easy to get through, was proposed in Einstein's 1905 article. This article's greatest achievement was to show how, once the Galilean transformations are replaced by Lorentzian ones, the above-mentioned principles compatibilize phenomena that appear disparate in different domains, such as mechanics and electromagnetism. The price to pay is merely to revise the Newtonian idea that there is an absolute time interval, independent of the coordinate system. In fact, that change was entailed by the necessity of revising Galilean transformations, since these are grounded on a classical interpretation of time and space. Here, the authors start to disagree – at least many philosophers do. In the following section, we introduce these authors' arguments. As it may seem, that which resembles relative simultaneity might in fact be absolute simultaneity, at least if we do not completely dispose of Newtonian concepts as, in Mach's words, 'metaphysical monstrosities' (Mach 1960).

2.1 *RoS* as a conventional result

A point often put forward by those who wish to reject *RoS* concerns the conventions assumed by the theory in the Einsteinian version (Craig 2008, 3, 21; Valentin 2008, 145; Salleri 2008, 180; Poincaré 1982; Reichenbach 1958; Mansouri and Sexl 1977, 497, 515, 809; Jammer 1979, 202-36). Einstein, in the 1905 article, ascribed physical meaning to a mensurable magnitude by making it conventional that any indication of simultaneity corresponds to 'that which is registered with clocks synchronized by beams of light'. This convention conceals some of its author's philosophical concerns stemming from anti-metaphysical doctrines with a strong verificationist flavour. Since this convention is based on a faulty epistemology that is thoroughly rejected in today, namely the verificationism empiricist, the result that hangs on it should also be revised accordingly. In other words, it suggests that *RoS* has a frail support, bound to an epistemology that is dated, faulty and false. As such, the results of conventional interpretations would be contaminated, considering that their acceptance is still owed to purely ideological reasons even today, or else for lack of knowledge of the relevant criticisms raised by epistemologists and philosophers to the verificationist meaning criteria (Craig 2002, 101-105; 2008, 37).[4]

---

[4]It might be laborious to find a single definition of 'empiricism' and 'verificationism'. However, the authors seem to accuse Einstein of adopting Mach's and others' acknowedgedly anti-metaphysical programme. Proponents of the latter were bent on eliminating anything that is non-observable from science. In this context, verificationism appears as a criterion of meaning: within the context of a scientific

Thus, it would be up to the mindful philosopher to warn the naïve scientist about the problems with this epistemological basis.

Einstein's endorsement of empiricism, especially the Machian kind, is fairly well known. It is not uncommon to ascribe a significant role to such concerns in the author's discoveries. Sklar precisely emphasizes the importance of empiricism for the author of relativity theory.

> Certainly the original arguments in favour of the relativistic viewpoint were rife with verificationist presuppositions about meaning, etc. And despite Einstein's later disavowal of the verificationist point of view, no one to my knowledge has provided and adequate account of the foundations of relativity which isn't verificationist in essence. (Sklar 1981, 141)

Other authors, among whom Reichenbach stands out, have always recognised this empiricist facet in Einstein's work. Experts know the role that the null results of the Michelson-Morley experiment could have had in the genesis of relativity, and the lesson that these result may have has a strong connection with Einstein's possible adherence to verificationism, adopted as a criterion of meaning and demarcation for scientific propositions. It is this empiricist commitment that determines, according to Reichenbach, what we can call Einstein's 'philosophy' (Reichenbach 1949: 291). Science historians, such as Michel Paty (1993), suggest that Einstein's empiricism is something known even among scientists who are accustomed to engaging in dialogue with the author's theories and especially among physicists who developed quantum mechanics in the 1920s. It is, therefore, an important facet of the author's thought that is historically based on Einstein's main philosophical influences (especially Mach and Hume).

Of course, the epistemology underlying the theory of relativity can be disputed. Today, although authors such as Craig (2002: 129-152; 2011: 16) consider the collapse of positivism – the kind of empiricism that Einstein supposedly adopted when accepting verificationism as a criterion of meaning – an opportunity to question the foundations of relativity, many authors would disagree with the premise that verificationist empiricism does not consist of a solid foundation. I believe that empiricism and verificationism are *prima facie* adequate as methodological principles, especially in the rigorous environment of the empirical and formal sciences. However, verificationist reasons are not

---

theory, only those theoretical 'objects' for which there is some method of verification have empirical meaning.

decisive in the domain of ontology, and denying the existence of entities, objects, relations or anything else only for verificationist reasons can be an unconvincing maneuver. The positivist association between the unverifiable and the non-existent is fragile, and this fragility is part of the strength of the *RoS* objector.

The mainstream argument points out that the conventions assumed upon defining 'simultaneity' are problematic because a) they assume that the speed of light is c in all inertial referentials, which is not proven within the theory itself; b) Newtonian absolute space is discarded, as well as other possible privileged referentials; c) the only justification for that is the impossibility of detecting absolute space, absolute movement or something equivalent (a purportedly verificationist justification).

As may be observed, the theory includes verificationism as a fundamental step in a supposition, at least according to the above-mentioned postulates and how they are articulated. Worse still, this supposition is significant for conventionally stipulating the meaning of 'simultaneity'. Einstein himself declared that we should understand this concept in a way presupposing that there is a 'physical meaning' (Einstein 1981, 393). The expression 'physical meaning' by itself is suggestive of verificationism. In the context of a scientific theory, the confirmation of a result should be its verification. Therefore, the next step is to show that the virtually unanimous rejection of verificationism compromises the most widely accepted interpretations of the theory. Einstein's Mach-inspired verificationist empiricism would contaminate the results of special relativity, such as *RoS* and time dilation. Not only is such verificationism obsolete, but it is also blatantly false according to most philosophers, at least as a criterion to reject the existence of something. But is it really that easy to defeat Einstein? Would the author of one of the most successful of our theories be such a naïve empiricist? This paper springs from the conviction that the complete story is yet to be told.[5]

---

[5]Naturally, the reasons for rejecting *RoS* also motivate the rejection of other relativistic effects, such as the 'length contraction' and the 'dilation of time', i.e., the said effects are then thought of as unreal, perspectival or false. My focus in this article is *RoS* for reasons that concern the result's philosophical interest. *Mutatis mutandis*, however, what is said in defence of *RoS* also applies to other relativistic effects.

3   *RoS* as a robust result

To defend *RoS* as a robust result, we do not need to deny the influence of empiricism on young Einstein's scientific and wider intellectual *Bildung*. Einstein's association with this epistemology is abundantly documented, making it unnecessary for our purposes to challenge what is firmly established. Our argument, however, implies that empiricism and verificationism are in fact a kind of Wittgenstein's ladder that Einstein 'climbed' with the purpose of questioning certain conceptions that had been accepted as true.[6]

It is not difficult to observe that the epistemology associated with special relativity plays a foundational role in the author's *Bildung*, leading him to suspect the results that were considered to be well established when they were in fact merely preconceived, without evidence or empirical support. This is also the case with the simultaneity of remote events, which we cannot actually observe. Among the authors who highlight this is Poincaré. According to him, there is no empirical evidence that supports the attribution of simultaneity to remote events (Poincaré 1982, 228). The indications of the simultaneity we have the capacity to know are local. As a rule, local simultaneities are merely coincidences between an indication of the clock's hand and an observable event, e.g., a train's arrival at a platform, the discharge of fireworks during a New Year's celebration, etc. However, how can we deny that *RoS* depends on verificationist premises, perhaps concealed, as explained in the former section? My hypothesis can be explained through a couple of fictitious principles:

> *Negative principle of existence attribution:* A theory should not acknowledge the existence of any 'object' which cannot be directly or indirectly verified, or whose rejection is a demonstrable consequence of directly or indirectly verifiable and accepted results.[7]

---

[6]In the Tractatus, Wittgenstein (2001, 6.54) says that a reader who has truly understood him, after arriving at a safe place, would abandon the 'ladder' used to reach that point. We may recover that image within our context, since Einstein, in adopting Machian empiricism, attains great achievements. The same Einstein, however, did not shy away from disregarding the empiricist Machian principles when it was necessary to obtain results (cf. Marques 2017, 12-13). According to Sklar (1981, 141), Einstein would come to refuse empiricism and verificationism later in his life, which makes 'Wittgenstein's ladder' an appropriate image.

[7]By 'object', we mean any theoretical result, e.g., that the speed of light should have the same constant value in any inertial frame, the existence of an ether, absolute space, etc.

Positive principle of existence attribution: A theory should acknowledge the existence of any 'object' that can be directly or indirectly verified, or whose acceptance is a demonstrable consequence of directly or indirectly verifiable and accepted results.[8]

While the negative principle is clearly verificationist, the positive one is not. However, both principles are empiricist and compatible with all we know about Einstein's intellectual *Bildung*. Our strategy is intended to depict the ultimate significance of the second principle: once shown to be applicable or compatible with *RoS*, we will not deem any of the relativistic effects to be dependent on that which was as a source of inspiration, namely, verificationism. Naturally, neither the principle nor any version of it was thought of by Einstein, although the first one resembles what some authors think has guided Einstein's verificationist reasoning. These are just general ways of condensing the reasons involved in the acceptance of theoretical results, such as the existence or non-existence of ether, absolute space or *RoS* itself. From our perspective, postulating fictitious principles might help discriminate the role of verificationism in that equation. The dialectics will be the following: the fact that non-verificationist principles are consistent with and help to explain the acceptance of the postulates that bear *RoS* shows that *RoS* is not necessarily verificationist. Naturally, this is consistent with the idea that verificationism acts as a 'Wittgenstein's ladder' in the theory's formulation.

According to the positive principle, the abundance of empirical evidence supporting the constancy of the velocity of light works as a sufficient reason to accept the postulate that the speed of light should have the same constant value in any inertial frame. Not including unobservable entities, e.g. ether, absolute space, could hardly count as a verificationist rejection of something. These principles should help clarify the difference between excluding entities and not postulating them. Perhaps we can understand the possibility of not proposing some entity, in the consecution of a new theory, without the accusation of verificationism.

To better understand our use of the above-mentioned principles, we may resort to a useful image: in a possible world $w1$ in which the neg-

---

[8]There are two observations regarding these principles. The first is that they are formulated by considering an attribution of existence by a theory, e.g. an ontological compromise that may be attributed to it. This is so, for as *tout court* principles for existence, both are compatible with false attributions. The second is that they are also formulated as meta-principles, to the extent that they also hold true for other principles that could be admitted.

ative methodological principle is adopted, Einstein may easily arrive at the same conclusions, postulating exactly the same conventions and premises stemming from the relativistic effects of the classical interpretation (some think that Einstein was guided by similar principles; hence, our world would be like $w1$ or at least close to it); however, there is a possible world $w2$ where Einstein adopts the positive principle without arriving at different results regarding what concerns the conclusions of either his theory or its postulates.

Imagine that a scientist puts forth the hypothesis that our universe is expanding, as suggested by Hubble's Constant.[9] Alternatively, imagine the hypothesis that light loses energy due to a dragging effect that is dependent on non-observable factors, such as a possible friction between light and the medium through which it is propagated, such as ether. Such a hypothesis (or a similar one) was proposed by Swiss astronomer Fritz Zwicky in 1929 (Zwicky 1929).[10] It is an alternative interpretation of the standard cosmological model. As is known, the standard model explains the same phenomenon in terms of a linear correlation between distances and velocities. In other words, the shift towards the red would evidence that the universe is expanding. Both solutions explain the phenomenon and both are equally valid interpretations of what was observed by American astronomer Edwin Hubble in 1929. However, Zwicky's hypothesis is merely of historical interest today.

We can apply both of these principles to explain the success of the hypothesis that the universe is expanding. However, only the positive principle will turn out to be appropriate, the reason being that one may claim that the absence of verifiable proof is not definitive proof that the rival theory is false. In other words, refusing an equally satisfying rival hypothesis is not something that can be justified by applying verificationist reasons. But since we can equally apply the positive principle mentioned, the claim that the standard cosmological model fails because it is contaminated by a faulty epistemology sounds extravagant. No verificationism is required to accept what experience, given the available information and knowledge, indicates. If an alter-

---

[9]The value of the Hubble constant is not yet accurately known. All measurements performed since the decade of the 2000s suggest a figure that varies between 63 and 73 km/s/Mpc.

[10]This is a version of Zwicky's theory, not its literal description. In the author's version, ether is not mentioned as a reason for the shift towards the red. Instead, only photons and very distant objects colliding are mentioned as the reason — the greater the distance, the bigger the collisions and the greater the shift towards the red. However, this detail does not alter the dialectic of our argument.

native hypothesis does not enjoy the same empirical precedent as its rival, it may be eliminated merely for being in conflict with the first.

The positive principle of existence attribution suggests a result that can help us to choose between rival interpretations of theories and does it in such a way that the selected interpretation is consistent with any choice guided by verificationist virtues, even in the absence of any verificationism.

3.1 The supposed equivalence between interpretations

The above defence against those who challenge the classical interpretations suggests that there are empirical reasons for preferring *RoS*. It would hardly be surprising to claim that the defiant authors, i.e., Markosian (2004), Quentin Smith (2008), Craig (2002, 2008), Valentin (2008), Maudlin (2008), Tooley (2008), Builder (1971), Ives (1979), Prokhovnik (1985, 1987, 1988), Balashov (2000), etc., do not believe in the decisive power of these reasons. Without them, the supporting empirical evidence does not justify the option of *RoS*. One might still point out that there is a fallacy involved in our comparison of both theories: there would be a disanalogy between the examples, since empirical evidence supports only the standard cosmological model, but not Zwicky's theory. However, this would be different in the case of special relativity, meaning that there would be supporting empirical evidence for both interpretations.

However, as explained above, no verificationist reason must be implied in the rejection of a rival interpretation. This is especially clear in the question raised by the shift towards the red and the standard cosmological model. The difference is that we know the effect through which light 'waves' change as the universe expands – the effect through which the length or frequency of a wave (an electromagnetic light wave) changes with the movement of a source relative to the observer – the so-called 'Doppler Effect'. There is, at least *prima facie*, an incompatibility between an expanding universe and a static one – that is, between a universe where the shift towards the red is the result of accelerating distant galaxies moving apart, and a universe where light loses energy quanta as a result of a dragging effect across huge distances. Hence, it seems that the solution to our problem would be to demonstrate that the same criterion is consistently applied to the theory of relativity.

After all, as is sometimes the case with metaphysical controversies, paralogisms of reason and philosophical aporiae, would the challengers of *RoS* be suggesting a full equivalence between rival interpretations? For instance, the question of whether the world is material – of the presence or absence of a 'substance' in the Lockean sense (cf. Locke 1999, 2. 23. 3) – could not be decided by inspecting matter or that which seems to correspond to the sensible *substratum* of the world. Would our challengers suggest something of this sort about the theory of relativity? Although we see the surrounding matter, it does not prove its existence, as Bishop Berkeley was well aware. Likewise, we conclude and observe results suggestive of *RoS* through the tests performed within the context of the theory, although they could be modified to challenge the theory. In the absence of something like a Lockean *substratum*, we would not know it, as Berkeley suggested (cf. Berkeley 1975, 92). In such a case, the world would behave as if there was, at least in our eyes.[11] If the defiant authors put their money on this equivalence, what can we specifically do to defeat them?

At any rate, we should ensure that the disputing sides of the current debate on *RoS* share some points in common. Thus, both proponents of an interpretation of the relativity theory share a commonality in terms of the scientific method, the requirement for grounding reasons on evidence and the belief in the possibility of resolving disputes, even if provisionally, by appealing to the consistency between theory and evidence. Of course, none of this makes verificationism a desirable methodological principle.

We think that the motivation for the challengers of *RoS* lies in two main reasons: 1) the supposed dependency on verificationism for a conclusion favourable to *RoS* and 2) an empirical equivalence between the rival theses. If we wish to attain our goal, we must also address the second point. Let us begin by discussing the thesis of equivalence among the many versions of the theory. There are different ways of trying to capture what Einstein's rivals aim at in proclaiming the empirical equivalence between interpretations. Let us assume that, given the character of the objection, what it means is that if we consider the influence of Newtonian space, luminiferous ether, etc, the things that seem to suggest *RoS* would turn out to be different; from this difference, an absolute simultaneity can be inferred. In this case, Einstein's rivals suggest that verificationist reasons are responsible for eliminating some important variables from the equation, which is how points 1 and 2 mentioned above are related. If those variables were available, then

---

[11]In this context, the reference to Berkeley is not out of place. According to Craig, Einstein would have drawn an ontological conclusion from purely epistemic reasons, just as Berkeley did. In fact, Einstein would have rejected absolute space, following what could be considered a version of *Esse est Percipi* (To be is to be perceived), which is probably the most important claim in Berkeley's work (Craig 2008, 4).

RoS, as well as the other relativistic effects, would be reinterpreted as merely a seeming result, perhaps to be corrected by a metaphysical description of the world.

Some authors further suggest that Lorentzian interpretations are, in fact, empirically better positioned if we consider larger contexts, such as Bell's theorem (Bell 1964, 1987) or the many different interpretations of particle physics (cf. Craig 2008, 5; Baron & Miller 2018, chap. 4). My reasons to remain sceptic about this suggestion are: a) the non-existence of theories that unify all the domains of physics, which could have supported the application of a Lorentzian approach; b) the fact that an interpretation of a theory (De Broglie 1928; David Bohm 1952, etc.) does not exactly count as evidence for other interpretations in other domains (Einstein 1905); c) none of the facts alleged in the objections to *RoS* seem compatible with the way in which light behaves without any theoretical addition to what is observed, which is consistent with Einstein's conjectures about whether these are conventional results or postulates. Of course, theories suggesting the existence of absolute simultaneity will conflict with relativity. However, without confirmation that light behaves differently from what Einstein claimed, this conflict will persist even if one is willing to correct the theory with claims consistent with absolute simultaneity. In other words, there is inconsistency among theories (relativity and quantum mechanics), interpretations (Einstein and Lorentz), domains (physics of macro and micro universes) and observations (measures suggesting instantaneous causes and light as an invariant limit for the speed of causal signals). Therefore, before different theoretical domains – or even interpretations of theories belonging to other fields of physics – can be used to choose between the versions of relativity, these conflicts in science should be resolved to bring us unified physics.

The best opportunity for our challengers is that those interpretations are equivalent because experience cannot decide between them. Effectively, the point is not to reject what scientists do in the laboratory; it is just that what they do is incapable of distinguishing between absolute and relative time measures, for which we could have some *a priori* preference. That makes us think the following: Not only can relativistic effects be equally explained by rival versions of a theory, but available empirical evidence may also suggest the existence of an ether, of an absolute space or a privileged referential. If this were the case, most scientists accepting *RoS* would do so for a subjective preference, along with some ideological component involved in the process. However, could we really believe that most scientists would refuse, for purely subjective reasons, all the supporting evidence for the existence of ether, absolute space or anything that could, at last, rehabilitate absolute simultaneity? This is quite difficult to believe. In fact, the claim that the available empirical evidence does not discriminate between rival interpretations, favouring both equally, is a cause for some perplexity for this very reason.

Special relativity's history of success could not disappoint us. Relativity has been subjected to empirical tests that have confirmed that it is consistent with phenomena. In fact, relativity is the paradigm of scientific theory that has been repeatedly subjected to tests and approved (Bondi 1964: 168; Taylor 1975, preface). However, the implication of the rival interpretation does not seem to even be perceptible, although it can be postulated in the construction of alternatives or inferred in case an absolute referential is accepted. However, this will be the case only if one wishes to challenge what experience seems to prescribe. As is known, time dilation and the relativity of simultaneity must be considered in many current technologies, and without them these would be impossible: from GPS to television sets firing high-speed electrons, everything seems to assume Einstein's successes. One of the most notable examples of the theoretical success of relativity lies in assertively pointing out how the half-life of *muons*[12] may be affected by velocities close to that of light (approximately $0.9998c$). The results are very close to what the theory might foresee – muons having a much larger half-life in relativistic circumstances, as time dilation leads us to suppose and, with it, the relativity of simultaneity and the length contraction.

The defender of the rival interpretations will claim that these very successes are compatible with absolute simultaneity. They would simply reflect the impossibility of measuring an absolute time interval, which does not entail its non-existence at all. In the next passage, Markosian makes it clear to us by dividing relativity into a stronger (STR+) and weaker version (STR-) of the theory. The weaker version, compatible with much empirical evidence, is also conciliable with absolute simultaneity:

---

[12]The muon is an elementary particle similar to the electron, with an electric charge -1 , a spin of 1/2 and a much bigger mass (105.7 MeV/c2). It is classified as a lepton, just like the electron (mass of 0.511MeV/c2), the tau (mass of 1777.8 MeV/c2) and the three neutrinos. As is the case with other leptons, it is not believed that muon has any substructure, that is, it does not present any simple particles.

Cf: https://cosmic.lbl.gov/SKliewer/Cosmic_Rays/Muons.htm. Consulted on 22/05/2020.

Although I agree that there seems to be a great deal of empirical evidence supporting the theory, I think it is notable that the same empirical evidence supports STR- equally well. And since I believe there is good a priori evidence favouring STR- over STR+, I conclude that STR- is true and that STR+ is false (Markosian 2004, 75)

In this context, how fair is it to claim the empirical successes of relativity for both interpretations? It seems that without the charge of verificationism, this move loses its strength. The thesis of empirical equivalence is supported precisely by the possibility of saying that the observed facts can be explained without relative simultaneity, except that we do not do so because we have a faulty epistemology. This epistemology favours interpretations that include *RoS*. But what if this epistemology is not necessarily implicated? Proving that *RoS* does not depend on verificationism will immediately weaken the claim of empirical equivalence. As such, it would be incorrect to proclaim that the examples chosen among the theories – that proposed by Zwicky and the alternative interpretations of relativity – are not symmetrical. What seems to be relative simultaneity can only be mistaken for absolute simultaneity if absolute space, or some other referential, can be adopted for empirical reasons. Likewise, the shift towards the red – an observable result that is consistent with different interpretations – will only have empirically equivalent explanations if the loss of light quanta can be verified. However, neither of the things that favour the rival interpretations are available, which does not amount to a verificationist elimination of the proposals. Rather, asserting a certain result when experience seems to suggest the same is what is at stake in this context. We can reinterpret the results naturally *ad infinitum*. However, not doing so would hardly count as a verificationist omission of equally probable hypotheses.

After all, if we wish to explain what happens in Einstein's train, even if we adopt a Lorentzian interpretation, we shall have to describe relative simultaneity, at least as a seeming result.[13] Furthermore, we never observe absolute simultaneity. All we can observe, under certain conditions, is the occurrence of a simultaneity *tout court*. The world,

---

[13]'Einstein's train' refers to the famous thought experiment carried out countless times by Einstein. Its purpose is to show the following: Two persons, one of whom is moving relative to the other, will report different plans of simultaneity. Assuming the theory's postulates, the only possible conclusion will be that both observers are right – simultaneity between events being a relative phenomenon (cf. Einstein 1999, 23-30).

even if it is not the ultimate truth of things, seems to bear relative simultaneity, for it is the only thing that is wholly compatible with what we think we know about the electrodynamics of bodies in movement. Postulating an ether or something similar, as well as being arbitrary, seems to involve an attempt to legislate how the world should function by invoking *a priori* reasons. If the main problem here is the arbitrariness of some Einsteinian notions, the arbitrariness in the challenging interpretations would make their position self-defeating.[14] The theorist who endorses *RoS* will only accept the result recommended by our positive principle of existence attribution.

The interpretations competing with those that accept *RoS* could be rejected on the basis of their incompatibility with the accepted interpretations, without being proven to be false. Therefore, we shall not have to demonstrate that all other interpretations are false in order to reject or accept the theoretical results. For instance, if we choose the positive principle, we will only have to select what experience seems to show, being aware that such a result might be false if there is something else, such as absolute space, ether or something equivalent. However, not deciding on absolute space will not be a faulty verificationist choice.

It must be noted that the charge of verificationism has this strange component: the behaviour of things in this world that does not correspond to what was initially foreseen must be avoided by the inclusion of *ad hoc* entities when required, even if this behaviour is not observed. Worse still, whoever fails to do so is committing a mistake, probably for verificationist reasons. I believe it will not be difficult to shake off this heavy charge from our shoulders, or even from Einstein's shoulders.

3.2   The definition of simultaneity

One of the main points of criticism against Einstein concerns what is called the 'conventional definition' provided by the author in support of the *RoS*. This is the definition of simultaneity put forward in Einstein's 1905 article and modified in his later works, although preserving the essential concept: a method by which synchronized clocks indicate

---

[14]Einstein's conventions are acknowledgedly arbitrary. For instance, to define simultaneity so as to exclude absolute space. Nonetheless, there is sufficient evidence for some of the suppositions that are not proven within the theory, e.g., that the speed of light is c in every inertial referential. Likewise, any reference to absolute space as capable of interfering with the behaviour of light, or even clocks and rulers, also sounds arbitrary. But isn't their supposed arbitrariness precisely the problem with Einsteinian conventions?

simultaneity through electromagnetic signals.[15] Einstein's definitions provide a non-neutral understanding of what events might be considered as simultaneous. Moreover, they also provide a method that is capable of making its identification replicable, manageable and, in the words of the current verificationism, 'significant'. In other words, the definition provides a method for verifying whether two events are simultaneous or otherwise, in different systems of coordinates. The focus of the criticism of the authors of this study lies in conventionally stipulating something about the behaviour of light for verificationist reasons. Einstein's definitions merely presuppose a light signal of constant velocity that can synchronize clocks operating at a distance.[16] Effectively, by using clocks synchronized through this method, we will be able to know whether two distant events are simultaneous. Here is an adequate paraphrase of one of the Einsteinian definitions:

1 *Two distant events, p1 and p2, are simultaneous if and only if two clocks synchronized by electromagnetic signals mark, at the same time (in the same instant t), the temporal indication of its occurrence.*[17]

In this section, we seek to express two things: a) that *RoS* does not depend on the convention assumed by Einstein, although the method it provides does; and b) a definition of simultaneity that does not include the same conventional or verificationist component – one that perhaps is even complementary with Einstein's.

Consider the following definition of 'simultaneous':

2 *Two events, p1 and p2, are simultaneous if and only if the time*

---

[15]Between 1905 and 1914, Einstein reformulated his definition of simultaneity four times. The general strategy of many reformulations was to reduce distant simultaneity to local simultaneity, which may be observed with the use of synchronized clocks. In 1917, Einstein reformulated his definition in a very particular manner. On this occasion, instead of synchronized clocks, he used the notion of a 'middle point' between distant events, i.e., equidistant observers of light signals, to identify the events that are simultaneous in their coordinate systems (cf. Nussenzveig 1998, 184).

[16]This refers to the 'Einstein synchronization' (or 'Einstein-Poncaré synchronization'), expressed as $t3 = t1 + 1/2(t2 - t1) = 1/2(t1 + t2)$. This indicates that the reflecting time of a light signal equals half the sum of the times marked on the clock while this signal comes and goes.

[17]In simultaneous events, instant t will exhibit a coincidence in both clocks when we discount the time taken by light to communicate the occurrence of the said events, i.e., a clock signals the instant at which certain events take place and communicates the same occurrence in other clocks that are spatially separated through an electromagnetic signal. If the electromagnetic signal is invariant to ether, absolute space, etc., we may discount the time of that trajectory and verify whether the instant t of every clock coincides for each event.

interval between them is zero.

Such a definition is essentially correct because when we compare simultaneous events, even when indicated by Einstein's method, they will always be capable of satisfying our definition, i.e., the time interval between simultaneous events will be equal to zero. However, relative to Einstein's definition, ours is neutral on many aspects – existence or non-existence of an absolute space, the behaviour and speed of light, whether or not *RoS* is valid, etc. Needless to say, it is because we accept that convention and assume that the behaviour of light is insensitive to the presence of ether – or absolute space – that we think it is possible to synchronize clocks by following the author's method. Moreover, as suggested by Einstein's definition, two events are simultaneous if and only if they can be indicated as such by synchronized clocks. In fact, another question presents itself to the Einstein's challengers. Even if Robertson-Walker's metric, microwave cosmic radiation, quantum vacuum electrodynamics, etc, could actually be indicated as a kind of ether or absolute space, it would still have to be demonstrated that light is sensitive to these referentials, resulting in a difference of velocity depending on the adopted coordinate system. An instance of this is when there is a difference resulting from a distortion in the devices used to perform a measurement.[18]

Einstein's definition, according to many, is an example of scientific conventionalism. Such a procedure that involves conventionally stipulating the aspects of a theory – e.g., meanings, operational definitions, etc. – might seem unjustified. However, it was not considered as such by many. Reichenbach, for instance, contrasts the role of axioms in physics – which should be as firm as they are in mathematics, i.e., they should be consistent, independent, unified, complete, etc. — with the role of definitions that are only arbitrary or conventional (Reichenbach 1924). According to Jammer, the arbitrariness of a definition indicates that the predicates of true or false do not apply to it (Jammer 2006, 175). Reichenbach further distinguishes between 'coordinative' and 'conceptual' definitions. The role of the latter is to clarify concepts by means of other concepts, while the former instills empirical

---

[18]This hypothesis, as known today, was reached by FitzGerald (1889) and Lorentz (1892) and became known as the FitzGerald-Lorentz contraction. This hypothesis, however, was still considered ad hoc in their time. According to the authors, the null effects of the Michelson-Morley experiment are the result of a 'contraction in the measurement instruments' and the variation of the interferometer is a supposed result of the movement of the Earth relative to ether, which would precisely have the value necessary to explain the results.

meaning using verifiable methods and is generally applied to primitive concepts, postulates or laws of a theory. However, independent of the role we ascribe to the conventions within a theory, we believe that there would be a consensus on considering that simultaneity can also be defined 'conceptually'. Additionally, the conceptual definition should agree with the attributions of simultaneity that the method contained in Einstein's definition might identify.

Definition 2, according to Reichenbach's distinction, is of the conceptual kind. It will probably not be useful for scientists, since it does not indicate the method through which we can identify simultaneity between distant events. This is also the method known to Einstein, which can be severed from the definition of simultaneity without prejudice. The advantage of our approach is merely that the conventional aspects of the theory are considered not as a matter of meaning but as a matter of method. What is made conventional is that the speed of light in vacuum is always c in any inertial coordinate system and, therefore, we can use electromagnetic signals with the purpose of knowing whether the interval between the clock's hand and distant events equals zero. In such cases, simultaneity will be achieved. Alternatively, if this does not happen, the events will not be simultaneous. To verify whether Einstein was wrong or right about the behaviour of light, we need to first inspect the world and then observe what it reveals to us. As far as anyone knows, there is no inconsistency between our best theories on the behaviour of light – an example being the null results from the experiments by Michelson and Morley (1887) – and the claim that light has the same speed in all inertial coordinate systems in a vacuum. If Einstein's interpretation of the behaviour of light is correct, despite the conventions that he made, the expected result will be RoS, independent of the method adopted to verify simultaneity.

It is still worth emphasizing that relativity has been formulated in many different ways in the last century, some of which do not use the postulate of the constancy of the speed of light or even make any important reference to the concept of 'simultaneity'. According to Norton (2004), the procedure based on clock synchronization by electromagnetic signals does not perform any fundamental role in the conclusion drawn for relative simultaneity. This suggests that Einstein's procedure is overrated from the point of view of those who reject the RoS. After all, one might conclude that the role of conventions is merely instrumental in implementing a theory.

4 Conclusion

We have assessed the claim that *RoS* has a built-in fault – a discredited and false epistemology. We also discussed other aspects of such criticism, such as the supposed empirical equivalence between interpretations that preserve *RoS* and those that reject it. Our objective has been to demonstrate that, according to the available evidence, the charge of verificationism would not hold true if the positive principle of existence attribution proves to be consistent with the theory's results. That is simply a way of clarifying how accepting experience and what it shows us, even when other hypotheses are available, is not a consequence of any verificationism, either implicit or explicit. After all, there is a noticeable difference between accepting P for empirical reasons and refusing its alternatives S for verificationist reasons. When P and S are conflicting explanations, the reasons for P might be vindicated without any verificationist rejection of S, as indicated by the historical example of the standard cosmological model versus its rival. This argument made use of fictitious principles, as well as comparisons between interpretations of theories, with the purpose of showing that we can present nonverificationist reasons to accept *RoS* even if empiricism has been a 'Wittgenstein's ladder' used by Einstein for a climb in which he discovered that time is not absolute, even if he later rejected the same ladder that allowed him to reach the coveted place.

Acknowledgments

This research was funded by the Portuguese Foundation for Science and Technology, grant number UIDB/00683/2020 (Center for Philosophical and Humanistic Studies, Universidade Católica Portuguesa). My Special thanks to professor Ricardo Santos who illuminated and improved my arguments by discussing a previous version of this paper. Thanks to Vitor Guerreiro for working on English in this paper. I also would like to express my gratitude to anonymous reviewers for their valuable comments on this paper, and to the Center of Philosophy of the University of Lisbon, for funding my participation in the Sixth International Conference on the Nature and Ontology of Spacetime. Special thanks to CNPq, Brazil.

References

Balashov, Y. 2000. "Enduring and Perduring Objects in Minkowski Space-Time." *Philosophical Studies* 99: 129166.

Bell, J. S. 1964. "On the Einstein Podolsky Rosen Paradox." *Physics* 1: 195-200.

Bell, J. S. 1987. *Speakable and Unspeakable in Quantum Mechanics.* Cambridge: Cambridge University Press.

Baron, S. and Miller, K. 2019. *An Introduction to the Philosophy of Time.* 1st ed. Cambridge. Polity Press, 280.

Berkeley, G. 1975. *Berkeley's Philosophical Works*, edited by Michael Ayers. London: Dent.

Bohm, D. 1952. "A Suggested Interpretation of the Quantum Theory in terms of "Hidden" Variables." *Physical Review* 85: 166-79.

Bondi, H. 1964. *Relativity and Common Sense.* New York: Dover Publications.

Builder, G. 1971. "The Constancy of the Velocity of Light." *Australian Journal of Physics* 11: 457-80.

Callender, C. 2008. Finding 'real' time in quantum mechanics. In: Craig, W, L. e Smith, Q., (ed.), *Einstein, Relativity and Absolute Simultaneity*, New York: Routledge. pp. 50-72.

Čapek, M. 1965. *El Impacto Filosófico de la Física Contemporánea.* Translated by Eduardo G. Ruiz. Madrid: Tecnos.

Craig, William L. 2002. "Relativity and the 'Elimination' of Absolute Time." In *Time, Reality, and Transcendence in a Racional Perspective*, edited by Peter Øhrstrøm, 99-128, Aalborg. Aalborg University Press.

Craig, William L. and Quentin Smith. 2008. "The Metaphysics of Special Relativity: Three Views." In Einstein, *Relativity and Absolute Simultaneity*, edited by William Lane Craig and Quentin Smith, 11-49, New York: Routledge. Callender, C. 2008. "Finding 'Real' Time in Quamtum Mechanics." In *Einstein, Relativity and Absolute Simultaneity*, edited by William Lane Craig and Quentin Smith, 50-72, New York: Routledge.

De Broglie, L. 1928. "La Nouvelle Dynamique des Quanta." In *Solvey* pp . 105-132. .

Dirac, P. A. M. 1951. Is There an Aether? *Nature* 168, 906-7.

Einstein. A. 1905. Zur Elektrodynamik bewegter Körper. *Annalen der Physik*, 17, 132-148.

Einstein, A. 1999. *A Teoria da Relatividade Especial e Geral.* Rio de Janeiro: Contraponto.

Fine, K. 2005. *Tense and Reality in Modality and Tense: Philosophical Papers.* Oxford: Oxford University Press.

Fitzgerald, G. F. 1889. *Science.* Washington, v. 13, p. 390.

Hubble, E. 1929. "A Relation between Distance and Radial Velocity among Extra-Galactic Nebulae." *Proceedings of the National Academy of Sciences of the United States of America* 15, no. 3: 168-173.

Ives, H, E. 1979. "Derivation of the Lorentz Transformations." *Philosophical Magazine* (1945) 36: 392-401, reprinted in *Speculations in Science and Technology* 2: 247, 255.

Jammer, M. 1979. "Some Fundamental Problems in the Special Theory of Relativity." In *Problems in The Foundations of Physics*, edited by G. Toraldo di Francia, 202-36. Bologna and North Holland, Amsterdam: Societa Italiana di Fisica.

Locke, J. 1999. *Ensaio Sobre o Entendimento Humano.* Lisboa: Fundação Calouste Gulbenkian. Original de 1689.

Lorentz, H. A. 1892. *Versl. K. Ak.* Amsterdam. Volume 1: 74.

Mach, E. 1960. *The Science of Mechanics: A Critical and Historical Account of Its Development.* La Salle: Open Court..

Marques, E. J. M. 2017. *Relationism, Empiricist Requirements and Relativity: Revista de Filosofia y Ciencias Prometeica.* Volume 0i15: 5-16.

Mansouri, R. and SexI, R.U. 1977. A test theory of special relativity: I. Simultaneity and Clock syncronization. *General Relatativity and Gravitation* 8: 497-513

Michelson, Albert, A. and Morley, Edward, W. 1887. The Relative Motion of the Earth and the Luminiferous Ether. *American Journal of Science* 34 (203): 333-345.

Maudlin, T. 2008. Non-local Correlations in Quantum Theory: How the Trick Might be Done. In *Einstein, Relativity and Absolute Simultaneity*, edited by William Lane Craig and Quentin Smith, 156-179, New York: Routledge.

Markosian, N. 2004. A defense of Presentism. *Oxford Studies in Metaphysics* 1, no. 3: 47-82. Minkowski, H. (1952). Space and Time. In *The Principle of Relativity: A Collection of Original Papers on the Special and General Theory of Relativity*, edited by H. A. Lorentz, Albert Einstein, H. Minkowski and H. Weyl, 75– 91, Dover, New York. .

Newton, I. 1990. *Principia: Princípios Matemáticos de Filosofia Natural.* Translated by Trieste Ricci. Sao Paulo: Ed. Nova Stella e Universidade de Sao Paulo. Original de 1687.

Norton, J. D. 2004. Einstein's investigations of Galilean covariant electrodynamics prior to 1905. *Archive for History of Exact Sciences*, 59 (1) pp. 45–105.

Nussenzveig, H, M. 1998. *Ótica, Relatividade, Física Quântica.* 1st ed., 437. São Paulo: Edgard Blucher.

Paty, M. 1993. *Einstein philosophe : La physique comme pratique philosophique.* Presses Universitaires de France.

Poincaré, H. 1982. The Measure of Time. In *The Foundations of Science*, translated by G. B. Halstead, 1913; , 228. University Press of America. Washington, D.C.:

Prokhovnik, S, J. 1985. *Ligth in Einstein's Universe.* Dordrecht: D. Reidel.

Prokhovnik, S, J. 1987. "The Twin Paradoxes of Special Relativity – Their Resolution and Implications." *Foudations of Physics* 19, no. 5: 541-52.

Prokhovnik, S, J. 1988. "The Logic of the Clock Paradox." Paper presented at the International Conference of the British Society for Philosophy of Science. *Physical Interpretations of Relativity Theory,* Imperial College of Science and Technology, London, 16-19 September.

Reichenbach, H. 1949. The philosophical significance of the theory of Relativity. In *Albert Einstein: Philosopher-Scientist*, P.A. Schilpp (ed.), The Library of Living Philosophers, Evanston. III., vol. 7: 287-311.

Reichenbach, H. 1958. *The Philosophy of Space & Time*. Dover: New York.

Sklar, L. 1981. Time, Reality, and relativity In *Reduction, Time and Reality*, edited by Richard Healey, 129-142. Cambridge: Cambridge University Press.

Salleri, F. 2008. "The Zero Acceleration Discontinuity and Absolute Simultaneity." In *Einstein, Relativity and Absolute Simultaneity*, edited by William Lane Craig and Quentin Smith, 180-211. New York: Routledge.

Smith, Q. 2008. "A Radical Rethinking of Quantum Gravity: Rejecting Einstein's Relativity and Unifying Bohmian Quantum Mechanics with a Bell-neo-Lorentzian Absolute Time, Space and Gravity." In *Einstein, Relativity and Absolute Simultaneity*, edited by William Lane Craig and Quentin Smith, 73-124. New York: Routledge.

Taylor, J. G. 1975. *Special Relativity*. Oxford Physics Series. Oxford: Clarendon Press.

Tooley, M. 2008. "A defense of absolute simultaneity." In *Einstein, Relativity and Absolute Simultaneity*, edited by William Lane Craig and Quentin Smith, 229-243. New York: Routledge.

Valentini, A. 2008. "Hidden Variables and the Large-scale Structure of Space-time." In *Einstein, Relativity and Absolute Simultaneity*, edited by William Lane Craig and Quentin Smith, 125- 155, New York: Routledge.

Wittgenstein, L. 2001. *Tractatus logico-philosophicus*. São Paulo: Edusp.

Zwicky, F. 1929. "On the Red Shift of Spectral Lines through Interstellar Space." *Proceedings of the National Academy of Science* 15, no. 10: 773-779.

Part II

The Ontological Nature of Spacetime

A. S. Stefanov, G. Dupuis-Mc Donald (Eds), *Spacetime Conference - 2022.*
Selected peer-reviewed papers presented at the Sixth International Conference on
the Nature and Ontology of Spacetime, 12 - 15 September 2022, Albena, Bulgaria
(Minkowski Institute Press, Montreal 2023). ISBN 978-1-989970-96-6 (softcover),
ISBN 978-1-989970-97-3 (ebook).

4 IS SPACETIME EMERGENT?

ANGUEL S. STEFANOV

Abstract   During the last two decades quantum gravity physicists have raised the claim that spacetime is an emergent entity out of a non-spatiotemporal reality at a quantum level. In this paper I present an alternative view. It is a "dissident" philosophical position in comparison to the emergentist doctrine about the nature of spacetime that is gaining popularity nowadays. My arguments are mainly ontological, and based on the idea that spacetime is a universal entity by itself.

Keywords: spacetime, quantum gravity models, reduction, gravitational waves.

1   Short Introduction

It seems that during the last two decades or so the claim about *the emergent character of spacetime*, not to speak of the three-dimensional space of our experience, has been getting support from the camp of physicists embarked on constructing models of quantum gravity. What is the reason for this claim, in spite of the fact that there is still no fully-fledged theory of quantum gravity to be accepted as such by the scientific community, and also having in mind that quantum field theory and the theory of general relativity (in the framework of which spacetime is a fundamental concept), are taken to be undoubtedly very well corroborated theories? The reason for this claim is the conviction that

> While spacetime is used in this theory [quantum field theory],
> it is not *described by* the theory. GR [general relativity], on
> the other hand, is a theory of spacetime. It describes spacetime
> itself as a dynamical field (that does not exist in some further
> 'background' spacetime, and so is *background independent*), and

A. S. Stefanov, G. Dupuis-Mc Donald (Eds), *Spacetime Conference - 2022.*
Selected peer-reviewed papers presented at the Sixth International Conference on
the Nature and Ontology of Spacetime, 12 - 15 September 2022, Albena, Bulgaria
(Minkowski Institute Press, Montreal 2023). ISBN 978-1-989970-96-6 (softcover),
ISBN 978-1-989970-97-3 (ebook).

says that gravity is due to the curvature of spacetime. Both of these theories are incredibly successful, yet, neither theory is thought to be fundamental... QG [quantum gravity] is supposed to be more fundamental than both these theories. [4, p. 4, author's italics]

As Crowther clearly states, the argument about spacetime emergence is that quantum gravity is supposed to be a more fundamental theory than general relativity and quantum field theory. Even if this is the case, however, this argument is prevailingly epistemological, not an ontological one. And I shall have it in mind when specifying my final conclusion (see section 5).

The aim of my paper is to present an alternative view. It is a "dissident" philosophical position in comparison to the emergentist doctrine about the nature of spacetime that is gaining popularity nowadays. My argument will be ontological, and based on the idea that spacetime is a universal and fundamental entity by itself. *A rather negative answer to the title question is reached at the end of the paper*, and what is meant by this phrase is also clarified there.

2 What is it for Spacetime to be Emergent, and not a Fundamental Entity?

One clear answer is that spacetime does not merely exist by itself, because it is emergent from something else – from some deeper non-spatiotemporal reality. This conceptual position is an eliminativist view of spacetime, since it denies its existence on a quantum scale. It is, however, a rather radical ontological view. It could be supported by quantum gravity physicists, but is reluctantly accepted by a philosopher with an ontological outlook like mine. I say this not because I cherish some conservative sympathy for spacetime, but because when denying the standard ontological status of spacetime one meets three conceptual difficulties.

The first one requires an explanation about our subjective experience not only about a flowing time (that has already been accepted by a lot of philosophers and scientists to be mind-dependent), but about our subjective experience of space, as well. As far as I am aware, an arguable explanation for the latter misrepresentation has not yet been suggested.

The second difficulty refers to carrying out experimental tests and empirical predictions of physical theories which are always accomplished at definite places and intervals of time. But quantum gravity theoretical schemes could not be tested in this way, and this raises a problem about their empirical confirmation.

The third difficulty runs to the fact that if spacetime is taken to be an emergent entity, then some conceptual approach must be contrived for the purpose of explaining what exactly is meant by the very notion of emergence in the case of spacetime, and not about any material system within spacetime. Another way of saying this is a reduction scheme to be exhibited, which could show how spacetime is reduced to a non-spatiotemporal entity, but such a clearly stated theoretical reduction is still not explicated. It could be objected, of course, that a putative reduction of an emergent entity to a more fundamental one is a too strong requirement for explaining the emergence as a posited fact. But yet some generic correlations must be definitely outlined, if one would like to take seriously the emergence of spacetime out of something else.

There are also voices against undertaking theoretical attempts at quantizing gravity. Such is, for instance, the clearly exposed consideration made by Antoine Tilloy [11] about the cogency of the alleged necessity for quantizing everything. I find his view to be convincing, but I shall not have it in mind here, since I'll stick to my own philosophical predilections.

Yet I may add that if one grasps gravitation not to be a physical force field, but as a result stemming out of the non-Euclidean geometry of spacetime, it may not be subject to quantization. For instance A. S. Eddington clearly mentioned this possibility, referring to gravitation as "supernatural agency" and saying that "gravitation as a separate agency becomes unnecessary". [5, p. 804]

Alongside the posited eliminativist doctrine there is a milder emergentist view that is usually known under the name of "*derivative spacetime view*". Since I find it to be the topical emergentist view nowadays, I dedicate my further analyses to this view. It states that *although spacetime exists, it does not exist fundamentally.*

Thus, two modes of existence are presupposed: one derivatively fundamental – that of spacetime, and another, "genuinely" fundamental – that of a non-spatiotemporal quantum basis. The derivative spacetime view pretends to show how a spacetime structure is *related to* the more fundamental quantum structure, or how spacetime qualities emerge from other kinds of entities. A paradigmatic and a widely spread construal in this respect is the functionalist one.

3 Functionalism

Functionalism is a legal step for saving a realist standpoint about the existence of spacetime. Many philosophers find a refuge in what space functionalism can provide. It is connected by default to a meaningful recovering of spatial features as we perceive them in human experience.

The conceptual backbone of functionalism is the claim that *spacetime is exactly what plays the role of spacetime in the framework of a quantum gravitational model.*

David Chalmers [3] and Sam Baron [1] are proponents of functionalism.

> As in the case of color, I think we've moved from primitivism to functionalism. We started with a kind of intuitive spatial primitivism where space involves these primitive qualities that we're acquainted with, and everything is spread out in that primitive space. We've ended with a spatial functionalism. To use a familiar functionalist slogan: space is as space does. Or better: space is whatever plays the space role. As with color, spatial properties need not themselves be functional properties (...), but they are picked out as what plays the functional role. This is analogous to various other familiar sorts of functionalism where we pick out Xs as whatever plays the X role. Colors are whatever plays the color role. [3, p .6]

All this means that there are some real entities and relations at a microlevel that engender something playing the role of space in human experience. [7] And this is so, because spacetime itself is an emergent conceptual construct from the standpoint of theoretical approaches to quantum gravity. This is an easy claim to be stated, insofar as these theoretical approaches like loop quantum theory, causal set theory, configuration space realism (wave function realism), and others, describe quantized fields underlying the spatial characteristics of observable events. But how should the declared spacetime emergence *per se* be gathered?

As Sam Baron has shown this kind of accepted emergence *is not of a mereological type* [1]. To this effect functionalism is welcomed, all the more so since it is a good explanation about the appearance of mental states. However, as he has demonstrated, even this standard form of functionalism does not work, and what he calls "partial functionalism" has to be resorted to.

> To be clear, I am not positing an entity 'approximate spacetime'

and saying that this new entity exists. The idea, rather, is that while spacetime *as we normally conceive of it* does not exist, the relevant functions of spacetime are performed by the ontology of a theory of quantum gravity, and they are performed in such a way that we can speak loosely of spacetime's existence, even though such talk is strictly speaking false. [1, my italics]

This way or not, by a standard or a weakened construal, functionalism tries to recover spacetime "as we normally conceive of it", or space as it is experienced by us. There are even other sorts of functionalism, but as Le Bihan has pointed out, all of them are not in a position to resolve the explanatory gap of spacetime emergence.

However, philosophers debating spacetime emergence through the lens of functionalism should adopt a clear view on the ontological picture they are relying on, if only for the sake of clarity and consistency of their proposal. They must adopt either one of the three substantive views – an identity view, a derivative view or an eliminativist view – or a neutral form of analytic functionalism which remains completely silent about the ontological implications of spacetime emergence. [2]

My ontological position suggested in section 5 is an alternative to "analytic functionalism which remains completely silent about the ontological implications of spacetime emergence".

Chalmers is also not very content with his own analogy, otherwise widely extended in his work, between color perceptions and the experience of spatial characteristics (e.g., distances among objects). The reason is that the analogy is ostensibly backed by a phenomenal functional recovery of space. However, space and spacetime are elements of well corroborated classical and semiclassical theories; and although recently contended to be only *derivatively* fundamental, space hardly resembles the phenomenal nature of colors in an ontological perspective. Colors are not theoretical constructs of scientific theories, being accepted only as mental representations and not as really existing features of things in the world. While on the contrary, space is an absolute entity in the Newtonian picture of the natural world, and material objects are situated in the four-dimensional spacetime of Einstein's general relativity, which has not ceased to receive observational confirmations.

A space functionalist need not recover *only* phenomenal properties. She ought to explain how spatial qualities, for instance metrical properties, could emerge outside human experience. In other words, an ontologically minded emergence that was thrown away by the common functional methodology peeps behind it. Still more, as it was briefly mentioned, there are different theoretical models pretending to quantize gravity (or spacetime), and this fact represents an additional difficulty. And it is probably the need for attaining a lucid ontological emergence that prompts Chalmers to look along this direction:

> The hardest case is grounding a reduction of space with a non-phenomenal analysis of the pretheoretical concept. It may be that the roles in interaction and motion that I've been gesturing towards can be used to deliver something even closer to the manifest image of space, picking out underlying properties in a quantum-mechanical or string-theoretic world as the grounds of motion and interaction...*I take that to be an open question which is going to involve a lot of detailed work* in the philosophy of physics combined with philosophical and/or psychological analyses of our concepts. [3, pp. 17-18, my italics]

It can further be noticed that the functionalist idea that spacetime can be reduced to what plays its role is a clear denial of spacetime as an ontological object, since it suggests that spacetime is posited as another thing for what *really* fulfills its physical role.

Loop quantum gravity seems to be quite popular among quantum gravity theories. This theory exploits as basic theoretical constructs granular quantum fields as elementary *grains of matter*, and the web of their interactions.

> The webs, in turn, transform into each other in discrete leaps, described in the theory as structures called 'spinfoam'.

> The occurrence of these leaps draws the patterns that on a large scale appear to us like the smooth structure of spacetime. On a small scale, the theory describes a 'quantum spacetime' that is fluctuating, probabilistic and discrete. At this scale, there is only the frenzied swarming of quanta that appear and vanish. [10, p. 110, my italics]

Although Rovelli does not use, in this case, a specific philosophical vocabulary and does not speak of spacetime functionalism, he quite lucidly points to the spinfoam leaps as drawing "the patterns that on a large scale appear to us like the smooth structure of spacetime". In other words, he points to the very structure in loop theory *that plays the role of spacetime on a large scale.*

Unfortunately, this presentation of the "derivative fundamentality" of spacetime is not yet the last word of contemporary physics. Rovelli's intellectual honesty is expressed by the following words:

> Loop theory is not a 'unified theory of everything'. It doesn't even begin to claim that it's the ultimate theory of science. It's a theory made up of coherent but distinct parts. It seeks to be 'only' a coherent description of the world as we understand it so far. (Ibid: 108, his italics)

And also:

> Am I certain that this is the correct description of the world? I am not, but it is today the only coherent and complete way that I know of to think about the structure of spacetime without neglecting its quantum properties. [10, p. 112]

There is no need to say that these words can also be taken to refer to the other attempts at constructing a fully-fledged theory of quantum gravity.

4 A Remark about the Posited Existence of Non-Spatiotemporal Grains of Matter

The same expression, "grains of matter" for naming non-spatiotemporal elements on a small scale is not a metaphor used only by Carlo Rovelli in the light of his already broadly admitted merit of being "the poet of physics". George Musser for instance speaks about "primitive grains of matter", depicted in the quantum gravity ideological canvas, "that do not exist within space, but simply exist – and stringing them together to form space". [8, pp. 218-219]

My remark about the posited existence of the non-spatiotemporal grains of matter concerns the spatial brim over the boundary of which space ceases its "derivatively fundamental" existence. Or, in other words, what is the spatial limit of the spatiotemporal (mode of) existence of spacetime as such? It is conjectured to be about the Planck length of 10^{-33} cm that equals 10^{-35} m. [10, p. 76]

Let me now turn to the recent announcements by the LIGO and Virgo Scientific Collaboration team from 2016 onwards about the registration of gravitational waves. What kind of waves are they? The accepted answer is that they are ripples in the spacetime fabric of the universe. These registered ripples, or waves within spacetime, could hardly be conceived of, if at distances having the average scale of their lengths spacetime would cease to exist. Their measured amplitude with strain is about 10^{-21} m. But spacetime is there to carry their permanently diminishing amplitudes from their cosmic source. They are longer than the Planck length, but yet represent minute spatial distances. Scientific teams are preparing to register gravitational waves with amplitudes of about 10^{-30} m, or high frequency waves, which stay closer to the ultimate edge of spacetime's non-existence, contended by proponents of quantum gravity models. (At that, we shall probably not have any experimental "confirmation" of spacetime's non-existence around the Planck length.)

Even a length of 10^{-18} m is 1000 times smaller than the diameter of a proton. This means that spacetime still really exists at extremely small distances before its physical reality to be *hypothetically* devoured by the grains of matter.

This remark clearly shows that spacetime is avowed to have a real physical existence on a very small scale, so we need not resort to Baron's rescue advice "that we can speak loosely of spacetime's existence" at least within the boundaries of our present experience. Concerning functionalism, *gravitational waves stay out of human perceptive abilities and their physical existence is accepted to be not controversial.* So, there is still no need for "foam leaps", or any other conjectured grains of matter to be taken *to appear for us* as space, if spatial distances are admitted to exist on such small scales. Thus, it seems that still there is no empirical urge for functional emergentism to recover spacetime out of a non-spatiotemporal reality.

If this reasoning is not convincing for the upholders of the emergent nature of spacetime, let me recall the posited basic elements of string theory, which is an alternative theory to encompass quantum gravity. These basic elements, the strings, are so minute to be estimated at 10^{-35} m in size, i.e., to be no larger than the Planck length itself. And yet, ontologically speaking, they produce different vibrational patterns to give birth to the different elementary particles. But where do they vibrate? In a ten-dimensional space is the answer. And the latter is an ontological entity that is wider than four-dimensional spacetime, and if it were taken to exist, then spacetime also exists as a part of it, and thus it is not properly emergent. To the objection that the strings are looked upon as still hypothetical basic elements of micro-reality, my answer is that they are at least components of a fully-fledged mathematical theory (with wide explanatory power), while the metaphorically suggested grains of matter are still not. Besides, string theory is a quantum gravity theory that does not presuppose non-spatiotemporal elements at the start, but relies on the quantum field hypothesis of the graviton to be responsible for the gravitational interaction.

5 My Answer to the Title Question

I said at the beginning of the chapter (section 1) that I shall give a rather negative answer to the title question. The answer is: "No, spacetime is rather not an emergent entity." I'll firstly give my ontological reason for the negative character of the answer, and secondly, I'll explain why I also decided to insert the specifying word "rather".

Like the proponents of the emergentist view of spacetime I admit a fundamental existence of the so-called "grains of matter", no matter what their accepted quantum qualities. Unlike them, however, I do not take their existence to be separated from the entity they underlie – the physical spacetime. By this claim I am not saying that the quantum grains of matter exist within spacetime. I merely say that they are the building bricks of spacetime. To this effect, they do not comprise a separate quantum base being an entity *ontologically distinct from spacetime*, while the latter appears only as an emergent side-product of them. On the contrary, the granular quantum base is a proper part of spacetime itself (or of a more dimensional space like that of the superstring theory).

By this claim I defend a methodological standpoint providing an interpretation of spacetime as an ontologically larger entity than is merely given by its standard geometrical presentation.

The following example is a pertinent one. Let us have in mind a definite volume of gas, e.g., ordinary air. Shall we say that the whole volume of air – as a thermodynamic entity – is not an entire entity, because it emerges out of its constituent molecules, and because its basic thermodynamic features like temperature and pressure are *derivable* from a theory referring to a lower level of structural description? No, we shall hardly say so. We shall certainly argue that *the volume of air is one self-identical entity*. And we shall do so notwithstanding that its constituents at a micro-level and its thermodynamic features at a macro-level *are described by two different theories*.

Well, my ontological claim is that we confront a similar situation concerning spacetime. Its deep quantum structure together with its higher-level metrical and topological qualities constitute one and the same entity – global spacetime. It is a fundamental entity by itself, notwithstanding the fact that different theories describe it on a large and on a small scale, as is the case with the corresponding theoretical presentations of the volume of air – a thermodynamical and a micro-structural one.

The adduced analogy with the volume of gas is illustrative for my metaphysical claim about the fundamental character of spacetime; so, it does not bear "a burden of proof", which is more desirable for the contenders of opposite emergentist views.

Yet another similar argument can be adduced in favor of my onto-logical contention about the global (or universal) nature of spacetime, one that is suggested by Brian Greene:

> Take a glass of water. Describing the water as a smooth, uni-form liquid is both useful and relevant on everyday scales, but it's an approximation that breaks down if we analyze the wa-ter with submicroscopic precision. On tiny scales, the smooth image gives way to a completely different framework of widely separated molecules and atoms. Similarly... Einstein's concep-tion of a smooth, gently curving, geometrical space and time, although powerful and accurate for describing the universe on large scales, breaks down if we analyze the universe at extremely short distance and time scales. Physicists believe that, as with water, the smooth portrayal of space and time is an approxi-mation that gives way to another, more fundamental framework when considered on ultramicroscopic scales. What that frame-work is – what constitutes the 'molecules' and 'atoms' of space and time – is a question currently being pursued with great vigor. It has yet to be resolved. [6, pp. 334-335]

Even if the last question obtains its future resolution, as is the expectation of many physicists, this would not be an obstruction to my claim that there are two theoretical approaches – on a large and on a small scale – to *one and the same entity*, be it a glass of water, or universal spacetime. To this effect (though probably intuitively), B. Greene writes about "molecules" and "atoms" *of* space and time, and not of a non-spatiotemporal reality. And this also stays in har-mony with R. Penrose's view that the quantum and the classical worlds should not be thought of as really being alien to one another. [9, p. 140]

A *rather negative* answer to the title question is a negative answer, of course, but with a specification. It does not change the logical sta-tus of the answer that concerns the ontological status of spacetime as a universal and fundamental entity. The specification indicates the *epis-temological need* for using different kinds of theories for the description of spacetime, at least for now – general relativity and quantum gravity theories. The latter are constructed with the hope of supplying a derivative explanation of the inherent geometrical qualities of spacetime. (Also, one may not forget the hope for a general theory that could encompass without conceptual problems all four types of interactions.) This is an epistemological reason for the chosen form of the answer (see Crowther's quotation at the beginning of the paper, in section 1). At that, we must also have in mind that another option was adumbrated by the string theory for both quantizing gravity and preserving spacetime.

References

[1] Baron, Sam. 2019. "The Curious Case of Spacetime Emergence." *Philosophical Studies*, https://doi.org/10.1007/s11098-019-01306-z, Springer.

[2] Bihan, Baptiste Le. 2019. "Spacetime Emergence in Quantum Gravity: Functionalism and the Hard Problem." *Synthese*, https://doi.org/10.1007/s11229019-02449-6, Springer.

[3] Chalmers, David J. 2020. "Finding Space in a Nonspatial World." In: B. Le Baptiste, N. Huggett, and C. Wüthrich (eds.) *Philosophy beyond Spacetime*. Oxford University Press. Forthcoming, PhilPapers.

[4] Crowther, Karen. 2021. "Spacetime Emergence: Collapsing the Distinction between Content and Context?" Forthcoming, In: Shyam Wuppuluri & Ian Stewart(eds.), *From Electrons to Elephants and Elections: Saga of Content and Context*, Springer.

[5] Eddington, A. S. 1921. "The Relativity of Time." *Nature*, 106, 17 February, 802-804.

[6] Greene, Brian. 2004. *The Fabric of the Cosmos. Space, Time, and the Texture of Reality*. New York: Alfred A. Knopf.

[7] Lam, Vincent and Christian Wüthrich. 2018. "Spacetime Is as Spacetime Does." *Studies in History and Philosophy of Science Part B: Studies in History and Philosophy of Modern Physics*, 64, 39-51.

[8] Musser, George. 2017. "Spacetime Is Doomed." In: Wuppuluri, Shyamand Giancarlo Ghirardi (eds.), *Space, Time and the Limits*

of Human Understanding. Springer International Publishing, 217-227.

[9] Penrose,Roger 2016. *Fashion, Faith, and Fantasy in the New Physics of the Universe*. Princeton University Press.

[10] Rovelli, Carlo. 2019. *The Order of Time*. Penguin Books.

[11] Tilloy, Antoine. 2018. "Binding Quantum Matter and Space-Time, Without Romanticism." *Foundations of Physics*, 48, Issue 12, 1753-1769.

5 Experiment: Mass does increase with velocity

Vesselin Petkov

Abstract   In the late eighties of last century an unprecedented assault on the concept of relativistic mass was launched mostly by some over-confident particle physicists, which, regretfully, still continues. It will probably go down in the history of physics as an unfortunate collective attempt to reject a concept firmly supported by experimental physics. Indeed, the relevant relativistic experimental evidence unambiguously demonstrates that *relativistic mass is an experimental fact*. Exactly like mass (defined as the measure of a particle's resistance to its acceleration), which reflects the experimental fact that a particle *resists* its acceleration, relativistic mass also reflects an experimental fact – the *increasing resistance* a particle offers when accelerated to velocities close to that of light; it is this increasing resistance that prevents a particle from reaching the velocity of light. It is demonstrated that the rejection of the relativistic velocity dependence of mass amounts to rejections of experimental facts and also to refusing to face and address one of the deepest open questions in fundamental physics – the origin and nature of the inertial resistance of a particle to its acceleration.

Keywords: mass, inertial mass, inertial resistance, relativistic mass

> This leads to a complete confirmation of the relativistic [mass] formula, which can thus be considered as experimentally verified.
>
> W. Pauli [1].

1 Introduction

During the last four decades physicists have endured "what has probably been the most vigorous campaign ever waged against the concept of relativistic mass"[1] [2].

---

[1] For a detailed account of the controversy over relativistic mass see Chapter 2 of Max Jammer's excellent book *Concepts of Mass in Contemporary Physics and*

A. S. Stefanov, G. Dupuis-Mc Donald (Eds), *Spacetime Conference - 2022*. *Selected peer-reviewed papers presented at the Sixth International Conference on the Nature and Ontology of Spacetime, 12 - 15 September 2022, Albena, Bulgaria* (Minkowski Institute Press, Montreal 2023). ISBN 978-1-989970-96-6 (softcover), ISBN 978-1-989970-97-3 (ebook).

It seems that campaign had been prompted by Adler's paper "Does mass really depend on velocity, dad?" [3] in which he had even discovered support for his denial of relativistic mass in Einstein's view[2] on this concept [3, p. 742]:

> Whatever Einstein's precise early views were on the subject, his view in later life appears clear. In a 1948 letter to Lincoln Barnett, he wrote

> "It is not good to introduce the concept of the mass $M = m/(1 - v^2/c^2)^{1/2}$ of a body for which no clear definition can be given. It is better to introduce no other mass than the 'rest mass' m. Instead of introducing M, it is better to mention the expression for the momentum and energy of a body in motion."

Unfortunately, Einstein's unclear view of relativistic mass[3] might have provided some encouragement for the campaign against the use of relativistic mass, but the above quote does not in fact demonstrate

---

Philosophy [2].

[2]Even the fact of seeking support for a physical argument not from the ultimate judge in physics – the experiment – but from Einstein's authority, appears to be a clear indication that that argument might be problematic. Moreover, Einstein himself would certainly have been glad if his authority had been questioned – once he said [4]:

> To punish me for my contempt for authority, Fate made me an authority myself.

[3]In his 1905 paper [5] Einstein defined two relativistic masses – longitudinal and transverse masses:

> Taking the ordinary point of view we now inquire as to the "longitudinal" and the "transverse" mass of the moving electron...Now if we call this force simply "the force acting upon the electron," and maintain the equation

> $$\text{mass} \times \text{acceleration} = \text{force}$$

> and if we also decide that the accelerations are to be measured in the stationary system K, we derive from the above equations

> $$\text{Longitudinal mass} \quad = \frac{m}{\left(\sqrt{1 - v^2/c^2}\right)^3}.$$

> $$\text{Transverse mass} \quad = \frac{m}{\sqrt{1 - v^2/c^2}}.$$

However, later Einstein avoided the entire concept of relativistic mass. See [6], [7].

that "his view in later life appears clear" – Einstein merely expresses his concern and reservation about the *definition* of M; this becomes evident when it is taken into account that the translation of the above part of Einstein's letter is inaccurate and misleading – compare the translation by Ruschin (there are no such phrases as "introduce no other mass than" and "Instead of introducing M") [8]:

> The German word *daneben* does not mean "instead of," but rather "besides," "in addition to" or "moreover." I would therefore translate the passage:
>
> It is not proper to speak of the mass $M = m/(1-v^2/c^2)^{1/2}$ of a moving body, because no clear definition can be given for M. It is preferable to restrict oneself to the "rest mass" m. Besides, one may well use the expression for momentum and energy when referring to the inertial behavior of rapidly moving bodies.[4]

Two years after Adler's paper L. B. Okun started a series of publications [9]-[15], which seem to had been the driving force behind the unprecedented campaign against the concept of relativistic mass. In May 1990 *Physics Today* published a number of letters to the Editor with comments on Okun's first article [9] and Okun's replies. W. Rindler's reaction was the sharpest [16]:

> I am disturbed by the harm that Lev Okun's earnest tirade (June 1989 page 31) against the use of relativistic mass ("It is our duty... to stop this process") might do to the teaching of relativity. It might suggest to some who have not thought these matters through that there are unresolved logical difficulties in elementary relativity or that if they use the quantity $m = \gamma m_0$ they commit some physical blunder, whereas in fact this entire ado is about terminology.

Unfortunately, after that exchange "Okun's polemic condemnation" [2, p. 53] even escalated – here are just two examples of his choice of words: "The pedagogical virus of relativistic mass" (from the abstract of a paper [12]) and "The Virus of Relativistic Mass in the Year of Physics" (a title of a paper published in the volume [13]). I think it is truly sad that such a prominent particle physicist did not seem to

---

[4]A scan of Einstein's letter in German is included in Okun's article [9].

have even attempted to entertain the possibility that he might have been fundamentally wrong.

While I share the feeling behind Rindler's reaction, I tend to disagree that "this entire ado is about terminology". And papers in support of the relativistic mass do show that the controversy implies more than terminology (see, for example, [17],[18]); here is the conclusion of Bickerstaff and Patsakos' paper [18, p. 66]:

> Thus we conclude by noting that in answering the elementary question of why two different masses are allowed in relativity, one obtains a clearer picture of the subject—a picture that is rooted in mathematics and logic rather than semantics and opinion.

Some physicists have argued that "There is no really good definition of mass" [19]-[22], which, according to them, might explain the relativistic mass controversy. This is simply untrue – since Newton mass has been defined as *the measure of the resistance a particle offers to its acceleration*. The attempts to reject the concept of relativistic mass appear to have been caused by the implicit use of the *everyday* (!?) understanding of mass – as the quantity of matter. It is worth reminding again Max Born's explicitly warning about the danger of improper understanding of mass in relativity [23]:

> In ordinary language the word *mass* denotes something like amount of substance or quantity of matter, these concepts themselves being defined no further... In physics, however, as we must very strongly emphasize, the word *mass* has no meaning other than... the measure of the resistance of a body to changes of velocity.

And indeed the accepted[5] definition of mass is precisely that – *the mass of a particle is the measure of the resistance the particle offers*

---

[5] Only several examples: "Mass is that property of an object that specifies how much resistance an object exhibits to changes in its velocity" [24]; "mass [is] the resistance of a body to a change of motion" [25]; Mass is "the quantitative or numerical measure of a body's inertia, that is of its resistance to being accelerated" [26]; "We use the term *mass* as a quantitative measure of inertia" [27, p. 9-1]; "Mass... measures how hard we have to push a body to achieve a given acceleration" [28]; "Mass is a quantitative measure of inertia... the greater its mass, the more a body "resists" being accelerated" [29]; "*The qualitative definition of the (inertial) mass of a particle is that it is a numerical measure of the reluctance of the particle to being accelerated*" [30]; "mass is a measure of the inertia of an object" [31]; Mass is defined as the "resistance to acceleration" [32]; "Mass is the measure of the gravitational and inertial properties of matter" [33].

to its acceleration. It is both adequate for the concept of mass in relativity and does *indisputably* demonstrate (see next Section) that mass indeed increases with velocity and therefore relativistic mass is an integral part of relativity (complementing proper or rest mass):[6]

A particle whose velocity increases and approaches the velocity of light *offers an increasing resistance to its acceleration*, that is, *obviously*, its mass (the measure of the resistance the particle offers to its acceleration) increases.[7]

This experimental fact (see next Section for discussing the experiments that established it) make the campaign against the concept of relativistic mass both inexplicable and worrisome. Instead of initiating and stimulating research on the origin of relativistic mass (and on the nature of mass in general) in order to achieve a more profound understanding of this fundamental concept in physics,[8] the relativistic mass is not mentioned at all in many publications[9] (see, for example,

---

[6]As I think it is exceedingly obvious that there are two masses in relativity (like two times; see below) – rest (or proper) mass and relativistic mass – I do not see any need to comment on the problems (coming from the equivalence of mass and energy) with a *single* concept of mass in relativity "as an invariant, intrinsic property of an object" [34].

[7]Arguing, effectively, that not the particle's mass but the particle's inertia increases [35] is not a real objection against relativistic mass. However, in reality, this is not a real objection, because inertia is the *phenomenon* of a particle's resistance to acceleration, whereas the particle's mass is the *measure* of that resistance. Therefore, if a particle's inertia, i.e. its resistance to its acceleration, increases, the measure of that resistance also increases; so the particle's mass increases.

[8]More research is needed to address the obvious situation: As the *resistance* of a particle to its acceleration depends on the acceleration's direction (the resistance is greater when the acceleration is along the particle's velocity and is becoming infinite as the particle's velocity is approaching the velocity of light), its mass is rather a tensor, not a scalar. In his 1905 paper [5] Einstein defined the two relativistic masses – longitudinal and transverse masses – but later silently abandoned them. With respect to the relativistic masses (longitudinal and transverse) we may witness a repetition of the story with the cosmological constant – initially Einstein used the cosmological constant in his equation linking matter and energy with the spacetime curvature, but later he called it the "biggest blunder of my life;" now cosmologists reintroduced Einstein's cosmological constant. At present time the relativistic mass (let alone the longitudinal and transverse masses) is so out of fashion that even such a prominent relativist as Wolfgang Rindler had to choose the words "confess" and "heuristic" in his letter to the Editor of *Physics Today* [16]: "I will confess to even occasionally using the heuristic concepts of longitudinal mass $\gamma^3 m_0$ and transverse mass γm_0 to predict how a particle will move in a given field of force."

[9]For a list of published works using relativistic mass see [36]; for more recent textbooks see, for example, [37]-[39]. Here I think it is worth mentioning specifically Feynman: "Mass is found to increase with velocity, but appreciable increases require velocities near that of light" [27].

the well-known textbook [40]) or, if it is mentioned, it is done to caution the readers[10], that "Most physicists prefer to consider the mass of a particle as fixed" [26, p. 760], that "Most physicists prefer to keep the concept of mass as an invariant, intrinsic property of an object" [34], that "We choose not to use relativistic mass, because it can be a misleading concept" [41] or to warn them [24, p. 1215]:

Watch Out for "Relativistic Mass"
Some older treatments of relativity maintained the conservation of momentum principle at high speeds by using a model in which a particle's mass increases with speed. You might still encounter this notion of "relativistic mass" in your outside reading, especially in older books. Be aware that this notion is no longer widely accepted; today, mass is considered as *invariant*, independent of speed. The mass of an object in all frames is considered to be the mass as measured by an observer at rest with respect to the object.

But phrases such as "*prefer* to consider," "*prefer* to keep," "*choose* not to use" (and "*can be*"), "*no longer widely accepted*" and even "older treatments" do not belong to the rigorous language of physics. Physics is not fashion where expressions such as "*prefer to*" and "*choose not to use*," for example, naturally fit. Physics at its best asks and addresses questions such as:

- Why is the velocity of light the greatest velocity, which cannot be reached by a particle possessing rest mass?

- Why does such a particle offer an *increasing resistance*[11] as its velocity increases and approaches the velocity of light? Or, which is the same question, why does the mass of a particle increase as its velocity increases and approaches the velocity of light?

---

[10]Some authors prefer to take a neutral position: "The use of relativistic mass has its supporters and detractors, some quite strong in their opinions. We will mostly deal with individual particles, so we will sidestep the controversy and use Eq. (37.27) $[\vec{p} = m\vec{v}\,(1 - v^2/c^2)^{-1/2}]$ as the generalized definition of momentum with m as a constant for each particle, independent of its state of motion" [29, p. 1244].

[11]In fact, the profound question of the nature of inertia and mass (i.e., the question of the *origin* of the *resistance* a particle offers to its acceleration) has been an open one since Galileo and Newton [42]. The discovery that mass increases with velocity and the controversy over relativistic mass made the need to try to address this open question more urgent.

2 Relativistic mass is an experimental fact

No matter how over-confident some physicists (mostly particle physicists) can be, the existing relativistic experimental evidence unambiguously demonstrates that the relativistic increase of mass is an *experimental fact* – again: it is an experimental fact that a particle offers an *increasing resistance* when its velocity increases and approaches the velocity of light and the measure of that *increasing* resistance is its increasing mass.

Let's start the discussion of the relativistic experimental evidence, which repeatedly confirmed the velocity dependence of mass, with Feynman's brilliant explanation of the role of relativistic mass in spacetime physics and its experimental confirmation[12] [27]:

> Newton's Second Law, which we have expressed by the equation
> $$F = d(mv)/dt,$$
> was stated with the tacit assumption that m is a constant, but we now know that this is not true, and that the mass of a body increases with velocity. In Einstein's corrected formula m has the value
> $$m = \frac{m_0}{\sqrt{1 - v^2/c^2}}, \qquad (15.1)$$
> where the "rest mass" m_0 represents the mass of a body that is not moving and c is the speed of light... Actually, the correctness of the [relativistic mass] formula has been amply confirmed by the observation of many kinds of particles, moving at speeds ranging up to practically the speed of light. ...
>
> What happens if a constant force acts on a body for a long time? In Newtonian mechanics the body keeps picking up speed until it goes faster than light. But this is impossible in relativistic mechanics. In relativity, the body keeps picking up, not speed, but momentum, which can continually increase because the mass is increasing. After a while there is practically no acceleration in the sense of a change of velocity, but the momentum continues to increase. Of

---

[12]As seen from his clear explanation one may reasonably wonder whether the campaign against relativistic mass would have been even launched, if Feynman had been alive at that time.

course, whenever a force produces very little change in the velocity of a body, we say that the body has a great deal of inertia, and that is exactly what our formula for relativistic mass says (see Eq. 15.10[13])—it says that the inertia is very great when v is nearly as great as c. As an example of this effect, to deflect the high-speed electrons in the synchrotron that is used here at Caltech, we need a magnetic field that is 2000 times stronger than would be expected on the basis of Newton's laws. In other words, the mass of the electrons in the synchrotron is 2000 times as great as their normal mass, and is as great as that of a proton! That m should be 2000 times m_0 means that $1 - v^2/c^2$ must be $1/4,000,000$, and that means that v differs from c by one part in 8,000,000, so the electrons are getting pretty close to the speed of light.

Feynman said it all:

- the concept of relativistic mass reflects something quite real – the *increasing resistance*, with which a particle opposes its acceleration when its velocity approaches c; that increasing resistance is the mechanism that prevents a particle from reaching c

- that increasing resistance is an experimental fact - "to deflect the high-speed electrons in the synchrotron that is used here at Caltech, we need a magnetic field that is 2000 times stronger than would be expected on the basis of Newton's laws."[14]

That is why, I think it is somewhat of a mystery how, after Feynman's comprehensive explanation, some physicists could reject the concept of relativistic mass, particularly particle physicists, given that the experimental evidence *proving* the relativistic increase of mass, provided by particle accelerators, is overwhelming – very powerful electric

---

[13]Eq. 15.10 reads $\mathbf{p} = m\mathbf{v} = m_0\mathbf{v}/\sqrt{1 - v^2/c^2}$.

[14]Now, even in (good) undergraduate physics textbooks (e.g. [44]) it is explained that "The predictions of the Special Theory of Relativity for the variation of the inertial mass with speed are taken into account in the construction of high-energy particle accelerators (synchrocyclotrons and synchrotrons) and are verified with accuracy up to very high energies (Sect. 7.3) [from Sect. 7.3: with conventional cyclotrons we achieve maximum energies only of the order of 5–25 MeV, lower than the theoretical limit. This happens because the condition of Eq. (7.27) presupposes that the mass m of each particle remains constant during the acceleration process. At relativistic speeds this is not true and m increases with the speed or the energy of the particles. The limit to the maximum energy that can be achieved with a cyclotron is due to this effect]."

and magnetic fields are used to accelerate electrons, protons and other charged particles. These particles need greater forces for their acceleration as their velocities approach the velocity of light c. Those greater and greater forces are needed *to overcome the increasing resistance* with which the particles oppose their acceleration as their velocities approach c. In other words, as the relativistic mass of the particles is the measure of the increasing resistance they offer to their acceleration, the greater forces are needed to accelerate particles whose masses relativistically increase.

But even the early experiments provided unambiguous experimental confirmation of the prediction of the 1905 Einstein's special relativity that the mass of a particle depends on its velocity[15] – for example, the 1908 experiment by the German physicist Alfred Heinrich Bucherer [45] (for a detailed discussion, see [46], Sec. 2.2 "Variation of mass with velocity") and the experiments by the Swiss physicist Charles Eugéne Guye and his collaborators (Ratnowsky and Lavanchy) [47]–[51]; for a detailed discussion, see [52] ("Replication of Guye and Lavanchy's experiment on the velocity dependency of inertia") and [53] (Sec. 9.2 "Experimental Verification of Relativistic Mass").[16]

These two experiments confirmed the velocity dependence of the electron mass. The proton relativistic mass variation was confirmed in 1958 [54]; here is the conclusion of the paper:

> The results are, within the limits of the experimental errors, in agreement with the relativistic law of the variation of mass.

To see why these first experiments did provide the decisive *proof* that the mass of a particle increases as its velocity approaches the

---

[15]The velocity dependence of mass was first predicted by the electron theory – that the electromagnetic mass of charged particles increases as their velocities increase – and generalized by Einstein for all particles: "these results as to the mass are also valid for ponderable material points" [5].

[16]Here is the conclusion of the 1916 Guye and Lavanchy paper summarizing the results of their experiment [50]:

> In summary, it emerges from the tables and the preceding graph, as well as from the considerations that we have just developed, that *the Lorentz-Einstein formula of the variation of inertia as a function of velocity is verified with great accuracy by all of our measurements.* [En résumé, il ressort des tableaux et du graphique qui précèdent, ainsi que des considérations que nous venons de développer, que *la formule de Lorentz-Einstein relative à la variation de l'inertie en fonction de la vitesse se trouve vérifiée avec une grande exactitude par l'ensemble de nos mesures.*]

velocity of light, I think a detailed explanation only of Bucherer's experiment is sufficient.[17] Rosser's [46, pp. 14-15] clearly demonstrates that:

In 1908 Bucherer measured the ratio of charge to mass (e/m) for β-ray electrons and showed that at high speeds, comparable to the speed of light, the masses of the electrons depended on the speeds of the electrons. ...

It can be seen[18] that the experimental values of e/m depend on the speeds of the electrons. However, if one assumes that

$$m = \frac{m_0}{\sqrt{1 - u^2/c^2}}, \tag{2.10}$$

where u is the speed of the β-ray electron and c is the speed of light, and if one calculates

$$\frac{e}{m_0} = \frac{e}{m\sqrt{1 - u^2/c^2}},$$

then the calculated values of e/m given in Table 2.1 are remarkably constant. They are as good a set of results as the reader is ever likely to obtain in his own laboratory work. In the spirit in which physical laws are 'established by experiment' in elementary practical courses, we will conclude from Bucherer's *experiment* that equation (2.10) is established by *experiment*. The quantity m in equation (2.10), ... is generally called the *relativistic* mass or just the *mass* of the particle. The quantity m_0 which is the value of m when $u = 0$, is called the *rest mass* or *proper mass* of the particle. It can be seen that as the velocity of the particle increases, according to equation (2.10) the mass of the particle increases.

Notice we assumed that the charge $-e$ on the electron was independent of its velocity. Instead of saying that mass varied according to equation (2.10), we might be tempted

---

[17]As Margenau put it[55]:

Even more direct is the observation of the *relativistic mass increase*, first accomplished by Bucherer and now a commonplace in every accelerator laboratory.

[18]From Table 2.1 of Bucherer's paper [45].

to say that the charge q on a particle varied according to the equation

$$q = q_0 \sqrt{1 - u^2/c^2}, \qquad (2.11)$$

where u was the velocity of the charge, q_0 the value of the charge when it was at rest, and that the mass m was invariant. Such assumptions would account for the results given in Table 2.1. There is, however, independent evidence in favour of the principle of constant electric charge. For example, if the charge on a particle did vary with velocity according to equation (2.11), then hydrogen atoms and molecules would not be electrically neutral, since the negative electrons are moving in orbits around the atomic nuclei in hydrogen atoms and molecules, and on average are moving faster than the positive nuclei (protons in this case) relative to the laboratory. If the charge did vary with velocity, hydrogen molecules should be deflected in electric fields, e.g. of the type shown in fig. 2.1 a. In 1960 King showed that the charges on the electrons and the protons in hydrogen molecules were numerically equal to within one part in 10^{20}. We therefore conclude that the charge on a particle is independent of its velocity and that the mass of a particle varies with the particle's velocity, according to equation (2.10).

The Bucherer experiment allowed *only* two interpretations and independent experiments ruled out the interpretation that the electron charge decreases as its velocity increases. So, it is exceedingly clear that this experiment would be *impossible* if the mass of electrons did not increase as their velocities increase. The same conclusion holds for all experimental evidence that confirmed the theoretical prediction that the mass of a particle depends on its velocity.

That is why rejecting the velocity dependence of mass amounts to rejecting experimental facts. This appears to have been clearly realized in 1921 by Pauli who wrote in his comprehensive (at that time) book *Theory of Relativity* (after discussing Bucherer's experiment) [1]:

The theory of spectra however gives us today, with the fine structure of the hydrogen lines, a much more accurate means for determining the velocity dependence of the electron mass. This leads to a complete confirmation of the relativistic formula, which can thus be considered as experimentally verified.

In a note added to the English translation of the book he cited the additional experimental evidence of the relativistic increase of mass obtained after 1921 [1, p. 216]

> Today the relativistic dependence of energy and momentum on velocity is taken as a matter of course in all experiments on high-energy particles, either occurring in cosmic rays or artificially produced with help of machines (cyclotrons, bevatrons, etc.) in which a high acceleration of charged particles takes place. For the computation of their orbits in these machines the relativistic formulae, the predictions of which have always been in agreement with experience, are also essential. A particular experiment, which *checked the relativistic mass formula* for electrons in a range of velocities up to nearly $0.8\,c$ was performed by M. M. Rogers, A. W. McReynolds and F. T. Rogers, Jr., *Phys. Rev.*, **57** (1940) 379. [italics added]

3 Addressing objections against relativistic mass

Some objections simply cannot be addressed, because, as shown in Section 1, they are not scientific objections – phrases of the type "*prefer to consider*," "*prefer to keep*," "*choose not to use*" and "*no longer widely accepted*" have nothing to do with the rigorous language of physics. For example, after an excellent introduction and discussion of the concept of relativistic mass, Allday writes [56]:

> While this is a perfectly valid approach to the issue of mass, it is not the only possible view.[19] Our colleagues in the particle physics community refer to rest mass as the mass of the particle, as the value is a characteristic property of the species concerned. For a particle in motion, particle physicists tend to work with the energy and momentum rather than worrying about the mass in such circumstances. We will follow their pattern.

In a note at the end of the chapter he explained the reason for not using relativistic mass [56, p. 136]:[20]

---

[19]The statement "it is not the only possible view" is just plain wrong as we saw in Sec. 2. For example, the detailed explanation of Bucherer's experiment unambiguously demonstrates that that experiment would be *impossible* if the electron mass were velocity independent.

[20]The dots at the end of the one-sentence note are Allday's.

I trained as a particle physicist and like to keep to the club rules...

I find such a reason amazing – should one assume that in physics "club rules" can overrule the verdict of the ultimate judge in science, particularly in physics – the experimental evidence?

Some physicists make general statements that the concept of relativistic mass might lead to confusions. For example, Weinberg [57] writes:

> Because of the presence of the factor γ $[\gamma = 1/\sqrt{1 - v^2/c^2}]$ in Eqs. (4.4.5) and (4.4.6), the quantity $m\gamma$ is sometimes called the relativistic mass. I will not use this terminology, because it suggests that we can calculate the acceleration produced by any force just by replacing m in Newton's $F = ma$ with $m\gamma$, which is not the case.

One may only wonder whether this can be regarded as a valid objection against the concept of relativistic mass, because it is merely not true that the relativistic mass "suggests that we can calculate the acceleration produced by any force just by replacing m in Newton's $F = ma$ with $m\gamma$"[21] and especially given the overwhelming experimental evidence proving that relativistic mass is an experimental fact. The situation is rather the opposite – it is the consistent use of relativistic mass which clearly demonstrates that it cannot merely replace the Newtonian mass in Newton's second law[22] as shown, for example, by Rosser [46, p. 16] while discussing "the changes in other mechanical quantities, arising from the variation of mass with velocity;" Feynman [27, p. 15-10] also discusses "some further consequences of relativistic change of mass." Moreover, Weinberg himself writes on p. 121:

> Given the existence of the force exerted by electric fields, the force exerted by magnetic fields is an inevitable consequence of Lorentz invariance. It is a special feature of electromagnetic forces that the only change in the equation of

---

[21] The concept of relativistic mass $m\gamma$ does not suggest anything. That a layman might get confused (and assume that one can simply insert $m\gamma$ in Newton's equations) is not a justification for not using $m\gamma$ in physics.

[22] In the general case, the relativistic force acting on a particle is not parallel to its acceleration and it also appears that the relativistic mass behaves as a tensor because a particle's resistance to its acceleration is *different* in different directions (an attempt to address this fact was already made by E. B. Rockower in 1987 [58]); it is greatest along the particle's velocity (serving as the mechanism that prevents a particle's velocity from exceeding that of light).

motion introduced by special relativity is the replacement of the mass m in the momentum with $m\gamma$, which in this one case allows us to treat $m\gamma$ as a relativistic mass.

Another such example is Giancoli's physics textbook [59]:

> In this "mass-increase" formula, m is referred to as the rest mass of the object. With this interpretation, the mass of an object *appears to increase as its speed increases.* But we must be careful in the use of relativistic mass. We cannot just plug it into formulas like $F = ma$ or $K = \frac{1}{2}mv^2$. For example, if we substitute it into $F = ma$, we obtain a formula that does not agree with experiment... In fact, many physicists believe an object has only one mass (its rest mass), and that it is only the momentum that increases with speed.[23]

A frequent, but physically irrelevant objection, is that rest mass is an invariant, whereas relativistic mass is the time component of the four-momentum. For example, this objection is the first part of the Taylor and Wheeler double objection against using the concept of relativistic mass [60]:

> The concept of 'relativistic mass' is subject to misunderstanding .... First, it applies the name mass – belonging to the magnitude of a 4-vector – to a very different concept, the time component of a 4-vector. Second, it makes increase of energy of an object with velocity or momentum appear to be connected with some change in internal structure of the object. In reality, the increase of energy with velocity originates not in the object but in the geometric properties of spacetime itself.

It is true that the magnitude of the four-momentum is proportional to the rest mass of a particle:

$$|\vec{p}| = mc.$$

The time component of the four-momentum

$$p^0 = \frac{mc}{(1 - v^2/c^2)^{1/2}} = m(v)c$$

---

[23] As Feynman explained it (see Sec. 2) the momentum increases due to relativistic mass increase.

is proportional to the relativistic mass $m(v) = m(1 - v^2/c^2)^{-1/2}$. So the rest (proper) mass m is indeed proportional to the magnitude of a four-vector and is an invariant, whereas the relativistic mass $m(v)$ is a component of a four-vector.

However, the situation is precisely the same with respect to proper time and coordinate time. The square of the spacetime distance Δs^2 between two events lying on a timelike worldline is equal to the scalar product $\Delta \vec{x} \cdot \Delta \vec{x}$ of the displacement four-vector $\Delta \vec{x}$ connecting the two events. In other words, the magnitude of the displacement vector is equal to the spacetime distance along the timelike worldline:

$$|\Delta \vec{x}| = \Delta s .$$

As $\Delta s = c \Delta \tau$, the magnitude of $\Delta \vec{x}$ is proportional to the proper time $\Delta \tau$ between the two events on the timelike worldline that are connected by the displacement vector:

$$|\Delta \vec{x}| = c \Delta \tau .$$

Therefore, the magnitude of the four-vector $\Delta \vec{x}$ is proportional to the proper time $\Delta \tau$.

On the other hand, however, coordinate time is the zeroth (time) component $\Delta x^0 = c \Delta t$ of the displacement four-vector $\Delta \vec{x}$.

So, if we cannot talk about relativistic mass, by the same argument we should talk only about proper time, which is an invariant, and deny the name 'time' to the coordinate time; however, it is the coordinate time that changes relativistically – the experimentally tested time dilation involves precisely coordinate time.

Therefore, proper or rest mass (which is an invariant) and relativistic mass (which is frame-dependent) are exactly like proper time (which is an invariant) and relativistic / coordinate time (which is frame-dependent) [and, to some extent, like proper and relativistic length].

It should be stressed that the resistance a particle offers to its acceleration (and therefore the increased resistance and energy) arises *in* the particle (more precisely, in the particle's worldtube); it does not come from the geometric properties of spacetime. It is spacetime that determines the shape of a geodesic worldline (and the shape of a geodesic worldtube in the case of a spatially extended particle), but it is the particle that *resists* when prevented from "following" a geodesic path, i.e., when the particle's worldtube is *deformed*.

We have proof that the resistance does not originate in the geometry of spacetime – a particle whose worldtube is deformed due to its deviation from its geodesic shape offers the *same* resistance in *both* flat and curved spacetime as the equivalence of inertial and passive gravitational masses shows (for more details see [61, Chap. 9]).

A somewhat relevant objection against relativistic mass appears to be the insistence that $\gamma = 1/\sqrt{1 - v^2/c^2}$ should not be "attached" to the mass, because it comes from the 4-velocity. That is, of course, correct [62] – γ ensures that the velocity of a particle cannot exceed that of light; in other words, γ ensures that no timelike 4-velocity vector, which represents the state of motion of a particle of non-zero rest mass, can become lightlike or spacelike. But that is kinematics; it says nothing about the dynamics, that is, it says nothing about why a particle cannot exceed the velocity of light (i.e. why the particle's 4-velocity cannot become lightlike or spacelike), and particularly – what is the mechanism that prevents it from doing so. That mechanism is, in fact, suggested by Newtonian mechanics, where mass is defined as the measure of the resistance a particle offers to its acceleration – when Einstein postulated that the velocity of light c is the greatest velocity a particle (with non-zero rest mass) can achieve, it was almost self-evident to assume that a particle would offer an increasing resistance when accelerated to velocities approaching that of light, that is, a particle's mass will increase and will approach infinity when the particle's velocity approaches c, thus preventing the particle from reaching and exceeding the velocity of light. And that was repeatedly experimentally confirmed. However, increased resistance and increased relativistic mass are rather only naming the mechanism that prevents a particle from reaching the velocity of light; the origin and nature of the resistance a particle offers when accelerated (an open question in classical physics) and of the increased resistance a particle offers when accelerated to velocities approaching that of light (an open question in spacetime physics) constitutes one of the deepest open questions in spacetime physics. Therefore, the rejection of the relativistic velocity dependence of mass amounts not only to rejections of experimental facts, but also to refusing to face and address an open question in fundamental physics.

4   Conclusion

Both mass and relativistic mass are equally supported by the experimental evidence—since mass is defined as the measure of the resistance a particle offers to its acceleration (which is the accepted definition based on the experimental evidence) and since it is also an experimental fact that a particle's resistance to its acceleration increases as the particle's velocity increases, it follows that relativistic mass also reflects an experimental fact—the increasing resistance a particle offers when accelerated to velocities close to that of light.

References

[1] W. Pauli, *Theory of Relativity* (Pergamon Press, New York 1958), p. 83

[2] M. Jammer, *Concepts of Mass in Contemporary Physics and Philosophy* (Princeton University Press, Princeton 2000) p. 51

[3] C. G. Adler, Does mass really depend on velocity, dad? *American Journal of Physics* **55**, 739 (1987)

[4] B. Hoffman, *Albert Einstein: Creator and Rebel* (Viking, New York 1972), p. 24

[5] A. Einstein, On The Electrodynamics Of Moving Bodies, *Annalen der Physik* **17** (1905): 891-921, in *The Collected Papers of Albert Einstein*, Volume 2: *The Swiss Years: Writings, 1900-1909* (Princeton University Press, Princeton 1989), pp. 140-171, p. 169

[6] E. Hecht, Einstein on mass and energy, *American Journal of Physics* **77**, 799 (2009)

[7] E. Hecht, Einstein Never Approved of Relativistic Mass, The Physics Teacher 47, 336 (2009)

[8] S. Ruschin, Putting to Rest Mass Misconceptions, *Physics Today*, May 1990, page 15

[9] L. B. Okun, The Concept of Mass, *Physics Today* **42**, 31 (1989)

[10] Л. Б. Окунь, Понятие массы, *Успехи физических наук*, Июль 1989 г. Том 158, вып. 3, с. 511-530 (L. B. Okun, The Concept of Mass, Usp. Fiz. Nauk. 158, 511 (1989) [Sov. Phys. Usp. 32, 629 (1989)])

[11] L. B. Okun, Putting to Rest Mass Misconceptions, *Physics Today* **43**, (1990) pp. 15, 115, 117

[12] L. B. Okun, The Concept of Mass in the Einstein Year, 2006, arXiv:hep-ph/0602037

[13] L. B. Okun, The Virus of Relativistic Mass in the Year of Physics, in: *Gribov Memorial Volume: Quarks, Hadrons, and Strong Interactions*, Proceedings of the Memorial Workshop Devoted to the 75th Birthday of V. N. Gribov, (World Scientific Publishing, Singapore 2009), pp. 470-473

[14] L. B. Okun, Mass versus relativistic and rest masses, *American Journal of Physics* **77**, 430 (2009)

[15] L. B. Okun, *Energy and Mass in Relativity Theory*, (World Scientific Publishing, Singapore 2009)

[16] W. Rindler, Putting to Rest Mass Misconceptions, *Physics Today*, May 1990, page 13

[17] T. R. Sandin, In defense of relativistic mass, *American Journal of Physics* **59**, 1032 (1991)

[18] R. P. Bickerstaff and G. Patsakos, Relativistic Generalization of Mass, *European Journal of Physics* **16**, 63–68 (1995)

[19] E. Hecht, There is no Really Good Definition of Mass, *The Physics Teacher* **44**, 40 (2006)

[20] E. Hecht, On Defining Mass, *The Physics Teacher* **49**, 40 (2011)

[21] A. Hobson, The Definition of Mass, *The Physics Teacher* **48**, 4 (2010)

[22] R. L. Coelho, On the Definition of Mass in Mechanics: Why is it so Difficult? *The Physics Teacher* **50**, 304 (2012)

[23] M. Born, *Einstein's Theory of Relativity* (Dover Publications, New York 1965), p. 33

[24] R. A. Serway, J. W. Jewett, Jr., *Physics for Scientists and Engineers with Modern Physics*, 9th ed. (Brooks/Cole, Cengage Learning, Boston 2014) p. 114

[25] W. Benenson, J. W. Harris, H. Stocker, H. Lutz (eds), *Handbook of Physics* (Springer, Heidelberg 2002) p. 37

[26] S. P. Parker (ed.), *McGraw-Hill Encyclopedia of Physics*, 2nd ed. (McGraw-Hill, New York 1993) p. 762

[27] *The Feynman Lectures on Physics. The New Millennium Edition.*
      Vol. 1 (Basic Books, New York 2010), p. 15.1, p. 15-9. See also
      Sec. 16-4 Relativistic mass

[28] M. W. McCall, *Classical Mechanics: From Newton to Einstein: A
      Modern Introduction*, 2nd ed. (Wiley, Chichester 2011) p. 5

[29] H. D. Young, R. A. Freedman, A. L. Ford, *Sears and Zeman-
      sky's University Physics with Modern Physics*, 13th ed. (Pearson
      Education, San Francisco 2012) p. 113

[30] D. Gregory, *Classical Mechanics: An Undergraduate Text* (Cam-
      bridge University Press, Cambridge 2006) p. 55

[31] D. C. Giancoli, *Physics: Principles with Applications*, 7th ed.
      (Pearson, New York 2014) p. 78

[32] W. M. Haynes, D. R. Lide, T. J. Bruno eds, *CRC Handbook of
      Chemistry and Physics*, 96th ed. (CRC Press, New York, 2015) p.
      2-58

[33] R. G. Lerner, G. L. Trigg, *Encyclopedia of Physics*, 2nd ed. (VCH
      Publishers, New York 1991) p. 703

[34] S. T. Thornton, A. Rex, *Modern Physics for Scientists and Engi-
      neers*, 4th ed. (Brooks/Cole, Cengage Learning, Boston 2013) p.
      61

[35] J. Roche, What is mass? *European Journal of Physics* **26** (2005)
      pp. 1-18

[36] G. Oas, On the Use of Relativistic Mass in Various Published
      Works, 2005, arXiv:physics/0504110 [physics.ed-ph]; see also
      G. Oas, On the Abuse and Use of Relativistic Mass, 2005,
      arXiv:physics/0504110 [physics.ed-ph]

[37] J. S. Walker, *Physics*, 5th ed (Pearson Addison-Wesley, 2017), p.
      1033 - "a constant force acting on an object generates less and less
      acceleration, $a = F/m$, as the speed of light is approached."

[38] R. D'Auria, M. Trigiante, *From Special Relativity to Feynman Di-
      agrams: A Course in Theoretical Particle Physics for Beginners*,
      2nd ed (Springer, Heidelberg 2017), p. 46: "*Given the new defi-
      nition of linear momentum, (2.7), the conservation of momentum
      is consistent with the principle of relativity, i.e. covariant under*

Lorentz transformations, if and only if the total relativistic mass is also conserved."

[39] D. Stauffer, H. E. Stanley, A. Lesne, *From Newton to Mandelbrot: A Primer in Theoretical Physics*, 3rd ed (Springer, Heidelberg 2017), p. 90: "The mass therefore becomes the greater, the greater the velocity v is, and becomes infinite when $v = c$."

[40] J. Walker, D. Halliday, R. Resnick, *Fundamentals of Physics Extended*, 10th ed. (John Wiley, New York 2014)

[41] K. S. Krane, *Modern Physics* 3rd ed. (Wiley, Chichester 2012) p. 51

[42] Newton explicitly defined inertia as the *resistance* a body offers to a change of its velocity (boldface added to "resisting" – V.P.): *"Inherent force of matter is the power of **resisting** by which every body, as far as it is able, perseveres in its state either of resting or of moving uniformly straight forward"* [43]

[43] I. Newton, *The Principia: Mathematical Principles of Natural Philosophy*, A new translation by I. Bernard Cohen and Anne Whitman assisted by Julia Budenz (University of California Press, Berkeley 1999) p. 404

[44] C. Christodoulides, *The Special Theory of Relativity: Foundations, Theory, Verification, Applications* (Springer, Heidelberg 2016), p. 314

[45] A. H. Bucherer, "Messungen an Becquerelstrahlen. Die experimentelle Bestätigung der Lorentz-Einsteinschen Theorie." (Measurements of Becquerel rays. The Experimental Confirmation of the Lorentz-Einstein Theory), *Physikalische Zeitschrift* 9 (22), pp. 755–762 (1908).

[46] W. G. V. Rosser, *Relativity and High Energy Physics* (Wykeham Publications, London 1969)

[47] C. E. Guye, Sur la valeur la plus probable du rapport e/m_0 de la charge à la masse de l'électron dans les rayons cathodiques, *Arch. Sc. Phys. et Nat.* (1906) 21: 461-468.

[48] C. E. Guye and S. Ratnowsky, Sur la variation de l'inertie de l'électron en fonction de la vitesse dans les rayons cathodiques et sur le principe de relativité. *Comptes Rendus de l'Académie* (1910) 150: 326-329.

[49] C. E. Guye and C. Lavanchy, Vérification de la formule de Lorentz-Einstein par les rayons cathodiques de grande vitesse. *Comptes Rendus de l'Académie* (1915) 161: 52-55.

[50] C. E. Guye and C. Lavanchy, Vérification expérimentale de la formule de Lorentz-Einstein. *Arch. Sc. Phys. et Nat.* (1916) 42: 286-299; 353-373; 441-448, p. 448

[51] C. E. Guye, Vérification expérimentale de la formule de Lorentz-Einstein. *Mem. Soc. Phys. Hist. Nat. Genéve* (1921) 39: 273-372.

[52] J. Lacki and Y. Karim, Replication of Guye and Lavanchy's experiment on the velocity dependency of inertia, *Arch. Sci.* (2005) 58: 159-170

[53] F. Rahaman, *The Special Theory of Relativity: A Mathematical Approach*, 2nd ed. (Springer Nature, Singapore 2022)

[54] V. P. Zrelov, A. A. Tiapkin, P. S. Farago, Measurements of the mass of 660 Mev protons. *Soviet Physics JETP*, Vol. 34 (7), No 3, pp. 384-387 (1958)

[55] H. Margenau, *Physics and Philosophy: Selected Essays* (D. Reidel Publishing Company, Dordrecht 1978), p. 197

[56] J. Allday, *Space-time, an introduction to Einstein's theory of gravity* (CRC Press, London 2019), p. 120

[57] S. Weinberg, *Foundations of Modern Physics* (Cambridge University Press, Cambridge 2021), p. 110

[58] E. B. Rockower, A relativistic mass tensor with geometric interpretation, *American Journal of Physics* **55**, (1987), pp. 70-77

[59] D. C. Giancoli, *Physics for Scientists and Engineers with Modern Physics* (Pearson 2014), p. 1118

[60] E. F. Taylor, J. A. Wheeler: *Spacetime Physics: Introduction to Special Relativity*, 2nd ed. (Freeman, New York 1992) pp. 250-251

[61] V. Petkov, *Relativity and the Nature of Spacetime*, 2nd ed. (Springer, Heidelberg 2009) p. 115

[62] V. Petkov, *Seven Fundamental Concepts in Spacetime Physics*, SpringerBriefs in Physics (Springer, Heidelberg 2021), Ch. 4

6 Dynamic Multipresentism: In Defence of a Dynamic View of Reality[1]

Jerzy Gołosz

Abstract The paper defends a dynamic view of reality termed dynamic multipresentism, which is founded on the assumption of the existence of the flow of time. This vindication makes use of a metaphysical theory of the flow of time developed by the author and is based on the notion of dynamic existence, which is a generalisation of becoming. It is demonstrated that such a conception allows one to explain the fundamental phenomena connected with the flow of time, namely the continuous changing of the present and the endurance of things. It is also shown that such a theory explains the origin and the asymmetry of time. It is argued that the proposed approach may be of some virtue for the empirical sciences because it explains the ubiquitous interest of scientists in the evolution of dynamic systems of different kinds and provides us with an arrow of time which is lacking in theories describing fundamental physical interactions: Causal Dynamical Triangulation is here analysed as a case study. The argument is advanced which aims to show that physics is unable to provide us with a theory of the flow of time and that we should look for such a theory in metaphysics.

Keywords: flow of time; change; dynamic existence; evolution; dynamic reality; dynamic multipresentism; origin of time, asymmetry of time; Causal Dynamical Triangulation

1 Introduction: The Problem with Dynamics and the Flow of Time

The paper defends a dynamic view of reality termed dynamic multipresentism, which is founded on the assumption of the existence of the flow of time. This vindication makes use of a metaphysical theory of

---

[1]The author would like to thank an anonymous reviewer for helpful comments.

A. S. Stefanov, G. Dupuis-Mc Donald (Eds), *Spacetime Conference - 2022.*
Selected peer-reviewed papers presented at the Sixth International Conference on the Nature and Ontology of Spacetime, 12 - 15 September 2022, Albena, Bulgaria (Minkowski Institute Press, Montreal 2023). ISBN 978-1-989970-96-6 (softcover), ISBN 978-1-989970-97-3 (ebook).

the flow of time developed by the author and is based on the notion of dynamic existence, which is a generalisation of becoming: it is shown that the notion of dynamic existence allows us to explain what the flow of time consists in and leads to dynamic multipresentism. It is also demonstrated that such a conception allows one to explain the fundamental phenomena connected with the flow of time, namely the continuous changing of the present, the asymmetry of time, and the endurance of things. What is more, it is shown that such a theory explains the origin of time itself.

Although the existence of the flow of time could seem to be obvious taking into account our everyday experience, we have so many conceptual difficulties with explaining the origin and mechanism of these dynamics, and especially those stemming from physics, that such a vindication is necessary.

There are some key phenomena of which our experience of the flow of time consists, and any plausible theory of the flow of time has to explain how they are possible and what their origin is. Namely:

I. The present is continuously changing: to put it metaphorically, it is "moving" toward the future;

II. We are convinced that while the present is "moving" towards the future, we and other things persist over time by enduring, that is, by being wholly present at each time at which we and other things exist and keeping strict (or literal, or numerical) identity;

III. We have traces of the past, and no traces of the future. The set of traces of the past both in our memory and around us is continuously growing;

IV. We can have an impact on future events without the possibility of an impact on the past;

V. The future seems to be open, while the past is fixed.

Now, the first and second phenomena seem to be especially important for our conviction about the existence of the flow of time because we feel – to say it metaphorically – "carried" by the flow of time into the future, keeping strict (or literal, or numerical) identity, and in the course of this process we are continuously changing in time. Space seems to be fundamentally different from time in this respect: even if there were some identical objects or parts of an object at the same moment of time, we would only say that they are *similar*, but we would not say that they are strictly identical to one another.

The third, fourth, and fifth phenomena form something which is called the asymmetry (or the arrow) *of* time (Sklar 1974, Chap. V), and could be easily explained by the flow of time (granted that we will have a plausible theory of the flow of time at our disposal).

In this paper, I will briefly analyse different aspects of the problem of the flow of time and attempt to show that the reasons for the difficulty with the acceptance of its reality may lie in inadequate metaphysics and a wrong approach to the relation between metaphysics and physics. First, in the second section, I intend to analyse this problem from the point of view of physics, and then, in the third section, from the point of view of philosophy, although it should be emphasised at this point that physical and philosophical contributions to our knowledge are so intertwined that the division of analysis between the second and third sections has to be somewhat conventional. What differs in these two sections is rather the dominant role of physics in the second section and philosophy in the third one. In the third section, I will present a metaphysical theory of the flow of time based on the notion of dynamic existence, which is a generalisation of the notion of becoming. It will be demonstrated that such a conception allows one to explain the fundamental phenomena connected with the flow of time, namely, the continuous changing of the present and the endurance of things. It will also be shown that this theory explains the origin and the asymmetry of time. In the fourth section, I would like to defend an approach to time and the relation between metaphysics and physics that can overcome difficulties blocking our understanding of reality as a dynamic one. In the same section, Causal Dynamical Triangulation is analysed as a case study.

2 Physics and the Flow of Time

If the idea of change and the idea of the flow of time in the world is to be saved, the challenges stemming from physics should be met. These are the following:

1. What is the shape of Now if

    a) there is no distinguished observer (frame of reference) in STR (special theory of relativity);

    b) there are no global hypersurfaces of *Now* in some models of GTR (general theory of relativity);

2. It is claimed that physics does not distinguish *Now*; physical laws are in force always (and everywhere);

3. It is claimed that physics is silent as it concerns the existence of the flow of time: there is no flow of time in physics and, especially, there is no theory of the flow of time in physics.

If the idea of the real flow of time is to be reconciled with physics, this short list should be completed by an analysis of the compatibility of the phenomena (II – V) with physics mentioned before in the *Introduction*, that is, by the analysis of the compatibility of endurance of things with physics (II) and the explanation of the fundamental asymmetry *of* time which is expressed in the points (III – V). In this way, we receive two additional challenges:

4. Is physics compatible with the endurance of things?;

5. The desired theory of the flow of time should explain the problem of the asymmetry of time, that is, first, the asymmetry of traces, secondly, why our actions can be only future directed, and thirdly, why the future is (or at least seems to be) open while the past is fixed.

Perhaps surprisingly, only the second question can be convincingly answered by physics itself: Quentin Smith showed that we should distinguish between *observational physics* and *theoretical physics*. For physical laws have to hold *always* and *everywhere*, and they cannot distinguish *Now*. That is why we should look for Now only in some *particular* physical processes, that is, in observational physics, not in physical laws. Thus, for example, we try to work out the *present* age of the universe, the *present* temperature of the cosmic microwave background radiation, the *present* value of the mean energy density in the universe, and so on (Smith 1985: 112-115).

However, contrary to what is widely believed, the fifth phenomenon, which I called after Sklar the asymmetry (or the arrow) *of* time, has not been explained by physics. The point is that attempts of explaining this asymmetry by invoking physical interactions directly or by means of a future directed causation hit upon a serious obstacle in the form of symmetry under the time reversal of the main physical interactions (strong, electromagnetic, and gravitational), which are supposed to be involved in the processes leading to the coming into existence of these traces and in our actions.[2]

---

[2] Weak interactions are not time reversal invariant but they are not involved in normal situations, for example, when we are reading, writing, eating etc. See e.g. Feynman (1967); Sklar (1974); and Gołosz (2017a, b; 2018: section V).

It should be emphasized that the thermodynamic explanation of the asymmetry of time is implausible as well because the process of the increase of entropy does not explain why we have traces of the past and no traces of the future; why we can have an impact on future events with no possibility of having an impact on the past; and why the future seems to be open, while the past is fixed and cannot be changed. An entropic increase cannot be responsible for the formation of traces and records because an increase in total entropy is not a necessary consequence of the formation of a record. Instead of this, entropy increase is necessary only for the *erasure* of a record (Hartle 2005: 106; Frisch 2010: Section 3; Gołosz 2021: 5). For example, if the wind wipes out the footprint-like traces in the sand increasing its entropy, it can be argued that one central role played by the thermodynamic arrow is to destroy macro-records and macro-traces rather than create them. As a consequence, we cannot consider the increase of entropy to be responsible for the asymmetry of traces. So, one can conclude that the increase of entropy as described by the second law of thermodynamics is only a process which is asymmetrical in time and in no way helps us to explain the asymmetry of time itself (see Sklar 1974, Chap. V; and Gołosz 2017b, 2021).

Nevertheless, in the next section, I would like to demonstrate, firstly, that philosophy has taken up the challenges $(1, 4, 5)$ successfully, and then I will attempt to show that the third, and perhaps most difficult, challenge can also be met.

3 Philosophy and the Flow of Time

There are convincing solutions to the first problem which have been proposed by philosophers and accepted by some physicists as well: to overcome these two difficulties, it is enough to choose the shape of *Now* in such a way that it will be *local*, and concomitantly it will look the same for all observers. As it transpires, however, it requires a change in our intuition about what is *Now*. Namely, after Stein (1968), Čapek (1976), and Shimony (1993), among others, we have to choose a *point-like present*; such a choice of *Now* satisfies both these conditions.

This view can be further developed: let us now assume that our world consists of things that are made up of approximately point-like objects and that *each of them* have approximately point-like presents as they crawl along their world lines. Such a point-like present can really be sensed by us, or rather by the cells and their different groups, of which we are made up at every moment of our life. Of course, a certain approximation is necessary, taking into account that the cells, although small, are nevertheless spatially extended and that the different cells of our nervous system need varying amounts of time to react and different amounts of time for the transmissions of signals. To avoid the objection that our organism and nervous system are spatially extended and as such cannot be treated in a point-like manner, they should be treated simply as collections of interacting parts – as small as one wants – each of which has its own individual time consisting of consecutive approximately point-like present *here-now*. To justify this approximation, I also recall at this point that the process of idealization is a normal procedure for all scientific research.

Such a *collective* approach to the point-like present, where approximately point-like interacting objects form greater collections such as an organism, provides us with an answer to the objection of solipsism according to which, first, a single point cannot be all that is real and, secondly, that it is not justified why an observer privileges his own *here-now* point over and above other observers' *here-now* points.[3] The answer is simple and *does not* privilege any observer: every approximately point-like thing constitutes its own approximately point-like present and interacts *directly in its own present* (that is, strictly speaking, in its own approximately point-like spacetime location) with different stimuli coming from other approximately point-like objects or with (quantum) fields, sources of which are other approximately point-like objects or their collections. So, according to the proposed conception, the world consists of approximately point-like interacting objects, each of which has its own approximately point-like present; such approximately point-like objects can form greater or lesser collections. This view can be termed *multipresentism*.[4]

One might certainly object that such an approach smuggles a global present by introducing multiple individual proper presents for all of the objects. However, such an objection would be invalid. To show this, it is enough to evoke an example of a physical model where we can say about the multiple individual proper point-like presents for all of the objects, although there is no global present in this model. In fact, we have many of such models, Gödel's (1949) model of a rotating universe being a primary example. In such a model, there are no global hypersurfaces of simultaneity and no other candidates for a global present, although we can obviously postulate that the smallest

---

[3]Such an objection is raised to standard idea of a point-like present of a *single* object/observer. See, for example, Saunders (2002: 286).

[4]A dynamics will be added to this conception later in this section.

constituents of these worlds, such as elementary particles, have their own approximately local point-like present (that is, their own approximately point-like spacetime location) as they crawl along their world lines. So, we have the multiple individual proper point-like presents with no global present in such a world, which shows that the former demands the latter in no way.

Contrary to what is sometimes claimed, there is no problem with the endurance of things in the context of physics (the fourth problem); the theory of relativity can be interpreted as a theory in which ontology consists of things as primary objects which keep their strict (diachronic) identity over time, and not as a theory in which ontology consists of events as primary objects. To show this, it is enough to point to the fact that we ascribe, for example, mass, momentum, energy, and rotation to particles or conglomerates of them and not to events.

The more serious difficulty is connected with the third problem, which has the form of the objection that there is no flow of time in physics and, especially, there is no theory of the flow of time in physics. I will split this problem into two subproblems, attempting to show, first, that metaphysics can provide us with a plausible theory of the flow of time which is able to meet the challenges $(1, 4, 5)$, and, second, that such solution can be acceptable for physics if only we agree that metaphysical hypotheses that have explanatory value for physics can be included in its research programs. This explanatory value would mean, *inter alia*, a solution to the problem of the asymmetry *of* time.

The most promising conceptions of the flow of time are based on the idea of becoming and have been developed by, among others, Bergson, Eddington, and Broad. Broad showed that becoming (which was termed *absolute becoming* by him) can be treated as a primitive notion so that the disastrous question "How fast does time flow?" can be avoided: "I do not suppose that so simple and fundamental a notion as that of absolute becoming can be analyzed, and I am quite certain that it cannot be analyzed in terms of a non-temporal copula and some kind of temporal predicate." (Broad 1938: 281).

Broad ascribed absolute becoming to instantaneous events, but if we – following Sellars[5] – ascribe it to things and additionally mix in with a conception of enduring, we receive a conception of dynamic existence of objects which can meet all the expectations related to the theory of the flow of time sought. This theory has the following form:

---

[5]"(...) whereas both *things* and *events* can become *Φ*, *only things can become in the sense of come into being*." (Sellars 1962: 556)

Dynamic Reality (DR): All of the objects that our world consists of exist dynamically, where the notion of *dynamic existence* is a primitive notion (just as Broad's absolute becoming) which can be roughly characterised by the set of postulates:[6]

i) the notion of dynamic existence is a primitive notion which is not based on any temporal predicate;

ii) things that dynamically exist endure;

iii) events (which are acts of acquiring, losing, or changing properties by dynamically existing things and their collections) dynamically exist in the sense of coming to pass.

Now, I would like to draw the attention to the essential fact, which can seem to be surprising, that when I introduced this conception of the flow of time based on the notion of dynamic existence, I *never* appealed to time itself; that is, I did not claim that *Now* is moving in time, nor that future objects and future moments of time are somewhere waiting to be fulfilled by dynamically existing objects. I could not have done this because it would lead to the difficult question of how fast time flows and to the four-dimensional block universe. What I have done instead is to introduce the notion of dynamic existence as a primitive notion which has the intrinsic property of directionality. This property means that the dynamic existence distinguishes one direction: toward the future.

But what about time? Where do *future* moments come from if they are not waiting somewhere to be fulfilled? I repeat that they cannot because it would mean introducing the four-dimensional static block universe. As it concerns this problem and my solution to it, I was inspired by Prior and his followers.[7] Namely, if the present is what exists, we can define it in a simple way:

The present ≡ The totality of objects that dynamically exist.

The present defined in such a way is continuously changing as things dynamically exist. And because we have added dynamics to existence,

---

[6]See Gołosz (2018: 403; 2020; 2021; 2022: *Introduction*; and 2023). The term "objects" is here used in such a way that it applies to things and events (facts and states of affairs), however things are treated as primary objects, while events as secondary.

[7]"The presentness of an event *is* just the event. The presentness of my lecturing, for instance, is just my lecturing" (Prior 1970: 247)

"To be present is simply to be, to exist, and to be present at a given time is just to exist at that time-no less and no more." (Christensen 1993: 168)

"On a presentist ontology, to exist temporally is to be present. Since presentness is identical with temporal existence (or occurrence) and existence is not a property, neither is presentness a property. Presentness is the act of temporal being." (Craig 1997: 37).

it almost automatically solves the problem of the *origin* of time: the dynamic existence of objects is responsible for the continuous emergence of new presents, and every new present means a new moment of time. So, according to this proposal, time is a consequence of the way we and other inhabitants of the world exist. That is, it is a *derivative of the dynamic existence* of objects: consecutive moments of time – each of which constitutes momentarily present — are *constituted* by dynamically existing objects, whatever they are. Time emerging in this way is a parameter which can be used to mark (or label, or measure) consecutive stages of the dynamic existence of objects. Due to the locality of the dynamic existence of objects and the locality of *now*, the time constituted by dynamically existing objects is their individual time, which can be equated with the so-called *proper time* of the theory of relativity.[8]

To finish this dynamic image of the world, we should supplement this conception by introducing the past and the future. They can be introduced by analogy to the earlier definition of the present:

The past ≡ The totality of objects that dynamically existed.
The future ≡ The totality of objects that will dynamically exist.

Now, it is easy to show that **DR** with the three definitions of the present, the past, and the future satisfies all of the posited conditions: firstly, because dynamic existence can be ascribed to the smallest constituents of the world, such as, for example, elementary particles, we receive a *local point-like present* for all of them, which is compatible with the theory of relativity. Secondly, it follows from the meaning of dynamic existence that things endure. Third, dynamic existence introduces an essential temporal asymmetry into the world because the past consists of things and events that dynamically existed and cannot be changed (at least some of objects from the past, such as, for example, Socrates, passed away while others, such as, for example, Donald Trump, still dynamically exist), while the future is to come into being and as such is probably open. The present is continuously changing as things dynamically exist. Due to this, despite the symmetry of strong, electromagnetic, and gravitational interactions under time reversal, things can transport traces of the past into the future, and they can only impact the present and future events. It would mean satisfying the fifth condition, which was imposed on the theory of the flow of time that we have been looking for.

It can also be seen that the stance introduced above deserves to

---

[8]See Gołosz (2015: 816 − 816; 2018 : 408, 411).

be called *dynamic multipresentism* because every present, which exists for every approximately *point-like* dynamically existing object along its world line, is dynamically changing.[9] And every approximately point-like dynamically existing object along its world line forms its point-like present in which it can interact *directly* with different stimuli coming from the past of other approximately point-like objects or with (quantum) fields, sources of which are others approximately point-like objects or their collections. That it is a form of presentism follows from the fact that, *by definition*, only the present exists. It does not lead to triviality, however, because this is **DR** that is the main ontological thesis of this view, while the sentence "**The present** ≡ The totality of objects that dynamically exist" is only a definition of the present.

Nevertheless, the question arises as to whether such a metaphysical theory of the flow of time can be acceptable for physics if it is supposed to be free of the flow of time? This is just the third challenge.

4 Final Remarks: Metaphysics as a Base for Physics?

First of all, I would like to notice that I cannot agree with Bergson that the empirical sciences (physics and biology, among others) offer us a static image of the world. To see this, it is enough to look at scientific theories and the image of the world they present. In fact, neither physics nor biology offers us a *theory* of the flow of time, nevertheless, they show us the world *in statu nascendi*, as a world which is dynamically changing. Thus, for example, physicists are especially interested in the evolution of the universe and other physical systems, and biologists are extremely interested in Darwin's theory of evolution. In all these cases, scientists are searching for theories describing the evolution in time of some dynamic systems, which would allow one to understand the mechanisms that underlie these processes and make predictions. We cannot find any theory of the flow of time in these theories, however, they describe the evolution of dynamic systems, and especially the evolution of the whole universe, which would be incomprehensible if the flow of time did not exist.

Therefore, if dynamic evolution is so strongly present in our world, an important question arises as to why these sciences, and physics especially, do not propose any theory of the passage of time. The answer does not seem too difficult: as was emphasized in the *Introduction*,

---

[9]See Gołosz (2023).

any plausible theory of the flow should explain two things: firstly, why this that exists, that is, the present objects, is continuously changing; and secondly, why we and other things persist through time by enduring keeping strict or numerical identity. This means that notions of existence, persistence through time, and diachronic identity over time should be involved in any plausible theory of the flow of time, that is, these notions and conceptions that are analyzed just by *metaphysics* and not by empirical sciences.

Instead of this, empirical sciences can analyze, for example, whether the elementary particles which make up dark matter exist and what their properties are, or how many dimensions space has, that is, *what exists* and *what properties* objects under considerations have, but not *what does it mean to exist* or whether these objects *keep diachronic strict identity or not*. Physics cannot decide whether the me of yesterday and me today are *one and the same person*. These last tasks seem to belong strictly to the domain of metaphysics, not physics.

In the antecedent section, I presented a metaphysical theory of the flow of time based on the notion of dynamic existence, which explains why the world is continuously changing and why we and other things persist through time by enduring. Due to the asymmetry introduced by the notion of dynamic existence, it also explains, as I tried to show, why we have an impact on future events with no possibility of an impact on the past; the asymmetry between the fixed past and the (probably) open future; and the asymmetry of traces: dynamic existence introduces an essential asymmetry into the world because the past consists of things and events that already dynamically existed and cannot be changed while the future will dynamically exist and as such is probably open. Things that dynamically exist endure toward the future (events which are acts of acquiring, losing, or changing properties by dynamically existing things and their collections dynamically exist in the sense of coming to pass) and due to this – despite the symmetry of strong, electromagnetic, and gravitational interactions under time reversal – they can only impact on things and events that happen at the same time or later, and can transport traces of the interactions into the future. In this way, the metaphysical conception of the flow of time presented not only explains the asymmetry of time, but also why empirical sciences, such as physics and biology, are so interested in the evolution of the world and its parts.

Finally, I would like to show that the proposed approach can help us develop modern physical theories as well, namely that dynamics and the temporal asymmetry introduced by the notion of dynamic existence can also be useful in developing theories of quantum gravity. This can be demonstrated by means of the example of Causal Dynamical Triangulation (CDT) developed by Jan Ambjørn, Jerzy Jurkiewicz, and Renate Loll. In this approach to quantum gravity, spacetime has a built-in arrow of time, which is used to distinguish between causes and effects, and, thanks to this, spacetime can be treated as emerging dynamically from causal time-asymmetric processes as a four-dimensional object.[10] It should be noticed, nonetheless, that although the arrow of time is necessary in this conception, this approach still does not explain what the *origin of the asymmetry (or arrow) of time* is. It is simply taken for granted that the temporal asymmetry is a primitive property of spacetime, which cannot be derived from a more fundamental theory. CDT also does not explain what the *origin of dynamics* is, which is necessary for four-dimensional spacetime to emerge: it again simply assumes that this dynamic in some way exists. However, *if we assume that there is a passage of time in the form of the dynamic existence of all of the objects that our world consists of, we receive both elements which are lacking in CDT: dynamics and the arrow of time.* Thus, CDT receives an interesting complement in the form of the proposed theory of time, which explains its foundation.

Summing up: I have tried to show that it is possible in a coherent way to defend the dynamic view of reality by means of the metaphysical theory of the flow of time, which is based on the notion of dynamic existence. Such an approach is in agreement with physics and explains:

- why the world is continuously changing;

- why we and other things persist through time by enduring;

- why we have an impact on future events with no possibility of an impact on the past;

- the asymmetry between the fixed past and the (probably) open future; and the asymmetry of traces;

- why empirical sciences, such as physics and biology, are so strongly interested in the evolution of the world and its parts;

- the origin of the proper time of the theory of relativity.

I deeply believe that the explanatory value of this metaphysical theory argues in favour of it.

---

[10]See Ambjørn, Görlich, Jurkiewicz, Loll (2008; 2014).

References

Ambjørn, J., Görlich, A., Jurkiewicz, J., Loll, R. 2008, "Planckian Birth of the Quantum de Sitter Universe", *Physical Review Letters*, 100: 091304, doi.org/10.1103/PhysRevLett.100.091304.

Ambjørn, J., Görlich, A., Jurkiewicz, J., Loll, R. 2014, "Quantum Gravity via Causal Dynamical Triangulation", in: Ashtekar, A., Petkov, V. (eds.) *Springer Handbook of Spacetime*, Springer, Dordrecht.

Broad, C. D. 1938, *Examination of McTaggart's Philosophy*, Cambridge University Press, Cambridge.

Čapek, M. 1976, „Inclusion of Becoming in the Physical World", in Čapek, M. (ed.), *The Concepts of Space and Time*, D. Reidel, Dordrecht.

Feynman, R. 1967, *The Character of Physical Law*; The MIT Press, Cambridge, MA.

Frisch, M. 2010, "Does a low-entropy constraint prevent us from influencing the past?", in *Time, Chance, and Reduction: Philosophical Aspects of Statistical Mechanics*, Hüttemann, A., Ernst, G., Eds. Cambridge University Press: Cambridge.

Gödel, K. 1949, "A Remark about the Relationship between Relativity Theory and Idealistic Philosophy", in Shilpp, P. A. (ed.), *Albert Einstein: Philosopher-Scientist*, Open Court, La Salle.

Gołosz, J. 2017a, "Weak Interactions: Asymmetry of Time or Asymmetry in Time?", *Journal for General Philosophy of Science*, 48:19 – 33, http://dx.doi.org/10.1007/s10838-016-9342-z.

Gołosz, J. 2017b, "The Asymmetry of Time: A Philosopher's Reflections", *Acta Physica Polonica* B, 48, No. 10: 1935 – 1946, http://dx.doi.org/10.5506/APhysPolB.48.1935.

Gołosz, J. 2018, "Presentism and the Notion of Existence", *Axiomathes*, 28: 395 – 417, DOI: http://dx.doi.org/10.1007/s10516-018-9373-7.

Gołosz, J. 2020, „In Defence of a Dynamic View of Reality", in Hanna, P. (ed.) *An Anthology of Philosophical Studies*, Vol. 14, Athens Institute for Education and Research, Athens, pp. 35 –47.

Gołosz, J. 2021, "Entropy and the Direction of Time", *Entropy*, 23 (4), 388; https://doi.org/10.3390/e23040388.

Gołosz, J. 2022, *In Defence of a Dynamic View of Reality*, Jagiellonian University Press, Kraków.

Gołosz, J. 2023, "How to Get out of the Labyrinth of Time? Lessons Drawn from Callender", *Logic and Logical Philosophy*, 32: 81–104, https://doi.org/10.12775/LLP.2022.015.

Hartle, J. 2005, "The physics of now", *Am. J. Phys.* 73, 101-109. doi: 10.1119/1.1783900.

Prior, A. 1970, "The Notion of the Present", *Studium Generale*, 23: 245–248, reprinted in *Metaphysics: The Big Questions*, Inwagen, P., Zimmerman, D.W. (eds.), Blackwell, Malden MA.

Saunders, S. 2002, "How relativity contradicts presentism", in Callender C. (ed.) *Time, reality & experience*, Cambridge University Press, Cambridge, MA.

Sellars, W. 1962, "Time and the World Order", in Feigl, H., Maxwell, M. (eds.), *Scientific Explanation, Space, and Time*, Minnesota Studies in the Philosophy of Science, University of Minnesota Press, Minneapolis.

Shimony, A. 1993, "The Transient Now", In *Search for a Naturalistic World*, vol. II, Cambridge University Press, Cambridge.

Sklar, L. 1974, *Space, Time, and Spacetime*, University of California Press, Berkeley.

Smith, Q. 1985, "The Mind-Independence of Temporal Becoming", *Philosophical Studies*, 47: 109–119.

Stein, H. 1968, "On Einstein-Minkowski Space-Time", *The Journal of Philosophy*, 65: 5 23.

Part III

INTERPRETATIONS OF RELATIVITY AND GRAVITATION

A. S. Stefanov, G. Dupuis-Mc Donald (Eds), *Spacetime Conference - 2022. Selected peer-reviewed papers presented at the Sixth International Conference on the Nature and Ontology of Spacetime, 12 - 15 September 2022, Albena, Bulgaria* (Minkowski Institute Press, Montreal 2023). ISBN 978-1-989970-96-6 (softcover), ISBN 978-1-989970-97-3 (ebook).

7 THE NATURE OF INERTIA, AND THE DYNAMIC GRAVITATIONAL FIELD

Branko Kovac

Abstract   The inertial force developed by accelerating the mass in the space-time continuum has no reaction force opposing it. Many scientists, including Einstein, recognized that as a serious problem. Presented is the hypothesis that the inertial force develops because the accelerating mass creates a force field around it in the same way as the weight of a mass is a consequence of the Newtonian gravitational field. The inertial force created by such a field has reaction force. The equation for the strength of the dynamic gravitational field produced around the accelerating mass is derived using the principle of action and reaction and the equation of motion $F = ma$. The dynamic gravitational field predicted by the equation is strong enough to be detected in the laboratory. This article describes the laboratory experiment which can prove or disprove the hypothesis of the dynamic gravitational field. The inertial force, calculated using the equation for the dynamic gravitational field, agrees with the behaviour of inertial force observed in the experiments on the Earth. The movement of the planets in our solar system calculated using that equation is the same as that calculated using Newton's method. The space properties calculated by the candidate equation explain the aberration of light and the results of light propagation experiments. The dynamic gravitational field can explain the discrepancy between the observed velocity of stars in the galaxy and those predicted by Newton's theory of gravitation without the need for the dark matter hypothesis.

Keywords: Gravitation; Gravitational fields; Non-standard theories of gravity; Laboratory experiment

1   Introduction

The idea that the mass accelerating in the inertial space creates inertial force can be traced back to Galileo and Newton. Galileo conducted ex-

A. S. Stefanov, G. Dupuis-Mc Donald (Eds), *Spacetime Conference - 2022.*
*Selected peer-reviewed papers presented at the Sixth International Conference on
the Nature and Ontology of Spacetime, 12 - 15 September 2022, Albena, Bulgaria*
(Minkowski Institute Press, Montreal 2023). ISBN 978-1-989970-96-6 (softcover),
ISBN 978-1-989970-97-3 (ebook).

periments by which he established that inertial force is proportional to the body's mass and acceleration relative to Earth. Newton developed a complete theory of mechanics where he introduced absolute space and time independent of physical phenomena. He used the concept of inertia, the property of the mass to resist change in speed, to explain the forces on the accelerating mass. When we accelerate mass in an absolute space, the accelerating mass creates inertial force because of that property. Einstein replaced Newton's concept of absolute space and time with the space-time continuum, but he kept the idea of the inertial property of the mass. When the mass, with its inertial property, accelerates in the local spacelike portion of the space-time continuum, it creates inertial force.

Einstein says about the concept of the space-time continuum [Einstein 2017 p. 59]: "As long as the principle of inertia is regarded as the keystone of physics, this standpoint is certainly the only one that is justified. But there are two serious criticisms of the ordinary conception. In the first place, it is contrary to the mode of thinking in science to conceive of a thing (the space-time continuum) which acts itself, but which cannot be acted upon." E. Mach criticized the concept of inertial force created in the absolute space or space-time continuum without any other mass. He advocated that mass is accelerated relative to the centre of all the masses in the universe. Mach has failed to provide the equations that would explain the influence that other masses have on the acceleration of one mass. The experiment has never been used to resolve the criticism that Einstein considers serious.

Some researchers attempted to provide a different concept for creating inertial force along the lines of Mach's thinking. One such attempt is by Sciama [Sciama 1953]. He used Maxwell-type equations to explain the propagation of the gravitational field, but as he pointed out, he was unsuccessful in providing a complete theory of inertia. Linearized equations of the general theory of relativity also produce a set of equations very similar to the Maxwell equations [Ruggiero & Tartaglia 2019]. Linearized equations can explain the propagation of the gravitational field, but the inertial force is still calculated as F = ma outside those force field equations.

In Galileo's and Newton's time, field theory did not exist, and neither of them could use it to explain the force that accelerating mass produce. Galileo assigned inertial property to the mass and explained inertial force by it. He has measured the acceleration from the fixed point on the Earth. Newton used the concept of absolute space, in which the masses move to be able to explain celestial mechanics. In preparation for understanding the new concept of inertial force based on the field theory, let's review the features of the force field that are common to electrical and gravitational forces:

- The object, either mass or charge, excites the space around it or creates a force field.

- Force on another object appears because that object is in the force field created by the first object.

- Forces appear in pairs and satisfy the principle of action and reaction. Both objects create the force field and the force of the same magnitude but in the opposite direction.

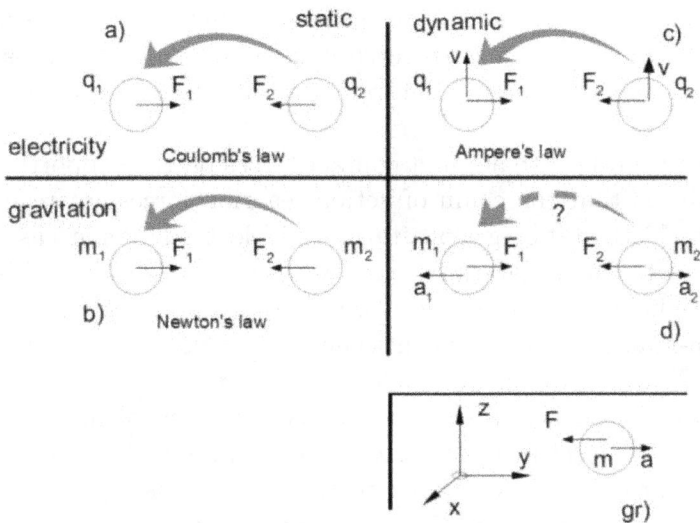

Fig. 1 Properties of force fields for electricity and gravity

The cases for creating electrostatic, gravitostatic, and electrodynamic forces are illustrated in parts a, b and c in Fig. 1. According to our understanding, the charge or mass excites the space around them. Another mass or charge feels the force because they are in that excited space. We call that space force field because the test charge or mass will feel the force when they are in that excited space. We see that both objects create a force field, and both objects create the force. The forces are opposite, and the principle of action and reaction is satisfied for those forces. We neglect the forces on supporting structures that are needed to hold static charge or mass in their position or keep moving charge on its course as a different mechanism of nature creates them. Fig. 1 part gr shows the creation of inertial force according to our current understanding. According to our current understanding, the inertial force is created because the inert mass accelerates in the local inertial space. Traditionally, the force by the supporting structure accelerating the mass is called acting force. If we neglect the acting force developed by the supporting structure, as it was neglected in previous cases, we see that the principle of action and reaction is not satisfied. To explain that the weight of the mass creates a reaction force, we don't look at contact forces beneath the mass. We look at the attraction forces that the mass creates on the Earth. In the same way, to explain the reaction to the inertial force, we shouldn't look at the contact force on the accelerated mass. Another mass that would create or feel reaction force is not present in creating the inertial force. Acceleration of mass in a universe without any other mass would still develop a force. The inertial force is created inside the mass in this case, while the force field in the space around the object creates a force in other cases.

When we take into account contact forces on the supporting structures, we see that the chain of action-reaction forces creates a closed loop for electrostatic, gravitostatic and electrodynamic cases. The chain goes through material bodies between two objects and is closed by a force field connecting those two objects through space. The chain of action-reaction forces is open when the inertial force is explained in the way Newton explains it.

This article explores the hypothesis that inertial force is created because every accelerating mass creates the gravitational force field in the space around it. According to this hypothesis, the inertial force develops because the mass is in the force field created by the other accelerating mass and not because of its own acceleration. Each mass should feel the forces of the same magnitude but in opposite directions; thus, the principle of action and reaction would be satisfied. The broken arrow in Fig. 1 d illustrates the hypothetical dynamic gravitational field.

Mach questioned whether the presence of other masses plays a collateral or fundamental role in mass dynamics [Mach 1989 p. 283]. Other masses are always present when we experiment with mass dynamics. We know that the presence of the Earth plays a fundamental role in creating the weight of the mass, so possibly the presence of the Earth also plays a fundamental role in developing the inertial force. Maybe the rotation of the water relative to the Earth causes water to recede from the centre of the bucket, in Newton's experiment, rather than the rotation in absolute space. The hypothesis of dynamic gravitational field considers the presence of the Earth to be fundamental in mass dynamics for the experiments carried out on the Earth.

The hypothesis of the dynamic gravitation field does not pose the question of the existence of separate gravitational and inertial mass. There is only one mass that creates two fields. One field is static, which mass at rest develops, and the other is dynamic, which the accelerating mass produces. This situation is similar to the case of electric charges, where we do not have separate static and dynamic charges. The open question is, does the dynamic gravitation field exist or not? That can be answered by the experiment.

The article shows that the dynamic gravitational field could explain the behaviour of inertial force on the Earth. The movement of the planets in our solar system can be explained by the field. The article looks at the consequences of this hypothesis on our concept of space. The hypothesis is consistent with the experimental and observational results related to light propagation on the Earth. The dynamic gravitational field hypothesis explains the velocity of stars in galaxies without the need for a dark matter hypothesis.

2 Estimating the strength of the dynamic gravitational field

The first question considered here is how strong should that hypothetical field be. To estimate the strength of the postulated field that creates inertial force, let's look at the forces on the mass we hold above the Earth, as illustrated in Fig. 2. The static force F_s, the weight of the mass, and dynamic force Fd, the inertial force, can be measured for the mass m that is held and accelerated above the Earth's surface. The forces of the same magnitude but in opposite directions should develop on the Earth.

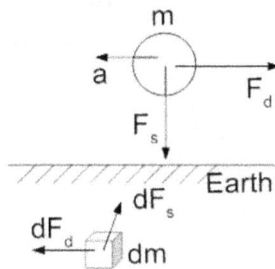

Fig. 2 Estimating the strength of the dynamic gravitational field

The small parcel of the Earth dm will have small reaction force dF_s created by the static gravitational field and dF_d caused by the dynamic gravitational field. The integral of the reaction forces dF_s over the whole of the Earth should be equal to the magnitude of the weight F_s of the mass m. The reaction force $-F_s$ on the Earth is calculated as

$$-\vec{F}_s = \int_V d\vec{F}_s = \int_V \vec{g}_s \, \rho \, dV, \qquad (1)$$

where $-F_s$ is the reaction force of the weight, g_s the Newtonian static gravitational field, ρ the Earth's density, and dV is infinitesimal volume.

In the same way, the reaction to the inertial force F_d can be calculated as a product of the mass dm and strength of the postulated dynamic gravitational field g_d produced by the mass m accelerated above the Earth. The reaction to the inertial force $-F_d$ can be calculated as:

$$-\vec{F}_d = \int_V d\vec{F}_d = \int_V \vec{g}_d \, \rho \, dV. \qquad (2)$$

If mass m is accelerated at about 10 ms^{-2}, the inertial force F_d and the weight of the mass F_s will be of the same magnitude. Then the magnitudes of the reaction forces dF_d and dF_s on the small mass of the Earth dm should be close to each other. The reaction forces dF_s and dF_d will not differ for several orders of magnitude. That means that the strength of the static and dynamic gravitational fields should be close.

The reaction forces calculated using the general theory of relativity are too small to account for the reaction to the inertial force. Einstein uses the word feeble to describe reaction forces on nearby masses developed according to the general theory of relativity. Frame dragging forces are so small that gravity probe B was needed to detect them [Everit et al. 2015]. If accelerating mass satisfies the principle of action and reaction, then the reaction force should come from a much stronger field than the gravitomagnetic field. Although it is much stronger than the gravitomagnetic field, this field is not strong enough to be noticed without the instrument.

3 The equation for the dynamic gravitational field

If it is not a gravitomagnetic field, the next question is, what does that field look like? To understand the hypothetical force field created by the accelerating mass and produce a candidate equation that can describe it, let us start with a simple system of two masses connected by the rod and the cylinder, as shown in Fig. 3. The assumption is that the rod and the cylinder are massless. If mass m_1 is accelerated by moving the rod inside the cylinder, mass m_2 will also accelerate in the opposite direction. The forces on mass m_1 and m_2 will balance each other.

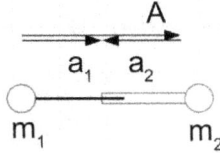

Fig. 3 System of two masses

The dynamic forces in the system are calculated using the equation of motion F = ma. The acceleration A between two masses can be measured. The individual acceleration can't be measured as we don't know the origin of the reference coordinate system. The difference in acceleration for individual masses a_1 and a_2 will equal the acceleration between masses A, as shown in Fig. 3. That difference can be used to calculate individual acceleration. For the system of two masses m_1 and m_2, which accelerate relative to each other, it can be written as:

$$-\vec{a}_2 = \frac{m_1}{m_2}\vec{a}_1 \qquad (3a)$$

$$\vec{a}_1 = \vec{A}\frac{m_2}{M}, \qquad (3b)$$

where M is the total mass $(m_1 + m_2)$, and other symbols are shown in Fig. 3.

To derive the expression for the dynamic gravitational field around accelerating mass, let's start with equation (3a), which represents the balance of forces in the system of two point masses. The mass m_1 will develop a dynamic gravitational field around mass m_2 that creates the force. In (3a), the negative acceleration for the mass m_2 is equivalent to the definition for the strength of the dynamic gravitational field developed by mass m_1 :

$$-\vec{a}_2 = \frac{\vec{F}_2}{m_2} = \vec{g}_{d1} = \frac{m_1}{m_2}\vec{a}_1 \qquad (4)$$

This equation is based on the balance of forces through the rod. The accelerations and the force are known from the experiments with mass dynamics. We already know the strength of the dynamic gravitational field at the location of mass m_2. Equation (4) should be modified to describe the dynamic gravitational field in space around the accelerating mass. We multiply each mass in equation (4) by the gravitational constant G and divide by the square of the distance between the masses r_{12} to get the following result for the dynamic gravitational field:

$$\vec{g}_{d1} = G\frac{m_1}{r_{12}^2}\frac{\vec{a}_1}{G\left(m_2/r_{12}^2\right)} = G\frac{m_1}{r_{12}^2}\frac{\vec{a}_1}{g_{s2}}, \tag{5}$$

where g_{d1} is the dynamic gravitational field developed by the mass m_1 at the location of mass m_2, G is the gravitational constant, a_1 is the acceleration of the mass m_1 relative to the centre of masses, r_{12} is the distance between masses m_1 and m_2, and g_{s2} is the magnitude of the static gravitational field of the point mass m_2 at the location of mass m_1.

Each mass in equation (4) is multiplied by G because the force field is gravitational. By doing so, the magnitude of the dynamic gravitational field can be related to the magnitude of the static gravitational field. Each mass is divided by the square of the distance because the field spreads from the mass evenly in every direction. The dynamic gravitational field is proportional to the mass creating it, and it decays with the square of the distance from the mass. The dynamic gravitational field is scaled by the ratio of acceleration and static gravitational field produced by the other mass in the system, namely mass m_2. The direction of the dynamic gravitational field is the direction of the acceleration of the mass.

Equation (5) describes the dynamic gravitational field created by the mass m_1 at the location of the mass m_2. That equation can be used to calculate the dynamic gravitational field anywhere in space using the distance between mass m_1 and the chosen point. The dynamic gravitational field created by the acceleration of the mass m_1 at any point in space is:

$$\vec{g}_{d1} = G\frac{m_1}{r^2}\frac{\vec{a}_1}{g_{s2}}. \tag{6}$$

Suppose several point masses are part of the system in addition to the masses m_1 and m_2. Their influence should be considered when calculating the dynamic gravitational field for the mass m_1. The masses will contribute to the magnitude of the static gravitation field at the location of mass m_1.

$$\vec{g}_{d1} = G\frac{m_1}{r^2}\frac{\vec{a}_1}{\sum_{i=2}^{n}G\left(m_i/r_{1i}^2\right)}. \tag{7}$$

136

It can be proven that the sum of the forces on all the other masses equals the reaction to the inertial force.

$$m_1 \vec{a}_1 = \sum_{j=2}^{n} m_j \vec{g}_{dj}. \tag{8}$$

The equation for the dynamic gravitational field with continuous mass distribution ρ for other masses in the system can be written as

$$\vec{g}_{d1} = G \frac{m_1}{r^2} \frac{\vec{a}_1}{\int_V G \left(\rho_i / r_{1i}^2 \right) dV}. \tag{9}$$

The volume integral in the denominator of equation (9) is a scalar quantity representing the magnitude of the static gravitational field from the other masses in the system. The dynamic gravitational field will be stronger for the same acceleration if the other masses in the system are smaller and further away.

Assuming the Earth is a sphere of radius R and uniform mass density ρ, the scalar static gravitational field for the Earth in the denominator of equation (9) is

$$g_s = 2\pi G \rho R \tag{10}$$

The vector static gravitational field on the surface of the Earth, according to Newton's law of gravitation, is

$$\vec{g}_s = -\frac{4}{3} \pi G \rho R \vec{r}_0 \tag{11}$$

The magnitude of the scalar static gravitational field is 50% higher than the magnitude of the vector static gravitational field for the Earth. The numerical value of scalar g_s at the Earth's surface is approximately $14.72 \ \text{ms}^{-2}$, so the dynamic gravitational field for the masses accelerated on Earth can be calculated using the following expression:

$$\vec{g}_{d1} \approx G \frac{m_1}{r^2} \frac{\vec{a}_1}{14.72} \tag{12}$$

The direction of the inertial force calculated this way will always be opposite to the acceleration of the mass. The dynamic gravitational field is in the direction of acceleration of the other mass in the system. The inertial force does not depend on the direction of acceleration relative to the Earth. It will be the same if we accelerate mass parallel or perpendicular to the Earth's surface. The inertial force is independent

137

of the location on the Earth. The inertial force will be the same for the same acceleration on the poles as on the equator. The reason for this is that the scalar static gravitation field of the Earth, and with it the dynamic gravitational field, changes with the location. Even when we move away from the Earth, the inertial force will always be proportional to the acceleration. The dynamic gravitational field is such that the principle of action and reaction is always satisfied. The inertial force on the accelerating mass will be matched by the reaction forces on the surrounding masses developed by the same physical process.

Mach's principle is present in the equation for the dynamic gravitation field. The inertial force is developed because the mass accelerates relative to the centre of all the masses in the universe. We can also estimate the error we introduce if we don't consider some of the masses when considering the inertial forces. Mach has failed to provide an equation that would mathematically define the inertial force by the field theory, while this article provides the equation. The equation can be tested by the experiment.

4   The motion of the planets and free fall

Let us check that the dynamic gravitational field equation agrees with the observed motion of the planets in our solar system. We need to verify that the acceleration of planets calculated using the equation for the dynamic gravitational field matches the acceleration of planets calculated using Newton's law of gravitation and equation of motion.

The equations for the balance of forces in the system with moving masses that have static and dynamic gravitational fields should take both fields into account. The forces produced by other masses' static and dynamic gravitational fields should always balance for each mass m_j in the system.

$$\sum_i \vec{F}_i = 0$$
$$\sum_i \vec{F}_{si} + \vec{F}_{di} = \sum_{i \neq j} m_j \left(\vec{g}_{sij} + \vec{g}_{dij} \right) = 0 \tag{13a}$$

In the above equation, g_{sij} is the static gravitational field created by the mass m_i at the location of mass m_j, and g_{dij} is the dynamic gravitational field of the mass m_i at the location of the mass m_j. The equation can be satisfied for any mass only if the sum of static and dynamic gravitational fields is zero. For the system of two point masses that have static and dynamic gravitational fields, the equations are:

$$G\frac{m_1}{r_{12}^2}\frac{\vec{a}_1}{G\left(m_2/r_{12}^2\right)} + G\frac{m_1}{r_{12}^2}\vec{r}_0 = 0$$
$$G\frac{m_2}{r_{12}^2}\frac{\vec{a}_2}{G\left(m_1/r_{12}^2\right)} - G\frac{m_2}{r_{12}^2}\vec{r}_0 = 0$$

(14a)

Applying the dynamic gravitational field equations to calculate the motion of a single planet and sun provides the same result for the planet's acceleration as applying Newton's law of gravitation and calculating acceleration using the equation of motion. That means that the orbital periods will be the same as if applying Newton's method. It can be seen that the dynamic gravitational field will be stronger, for the same acceleration, for the outer planets in the solar system than it is for the inner planets. Without the effect of scaling the acceleration with the static gravitational field, the reaction force on the sun would not balance the force of the static gravitational field of each planet.

The first difference when calculating the movement of the planets using the dynamic gravitational field and applying Newton's law of gravitation is that the acceleration of the planets would create a force on the other planets. The dynamic and static gravitational fields for each planet would influence the movement of the other planets. The second difference is that all planets and the sun should have acceleration. We can't have a system in which the sun is stationary. The acceleration of the planets and the sun need to be considered, or there will be an imbalance of forces. Both of these effects are already known. We know that the sun is orbiting around the solar system's barycentre and the sun's acceleration influences other planets' movement.

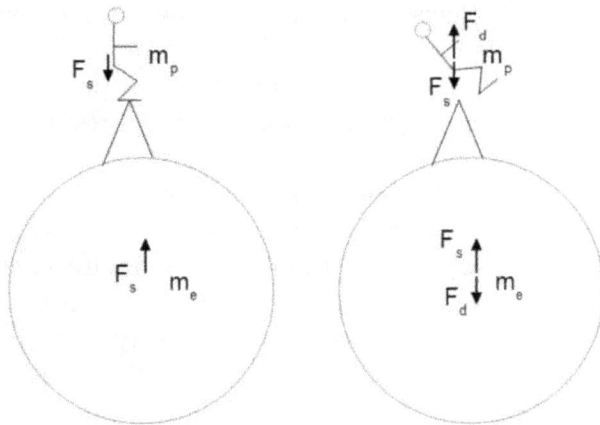

Fig. 4 Person on a ladder losing the footing

139

To review the forces during free fall, we can consider the person on a ladder losing the footing. On the left-hand side of Fig. 4, we see a person standing on a ladder. The person will feel the force of the weight F_s on the contact point with the ladder. The person's weight is matched on the Earth by force $-F_s$ with which the person attracts the Earth. On the right-hand side, the person loses their footing and falls. When the person starts free fall, there will be no external force acting on the person. The weight F_s will be balanced by the inertial force F_d. At the same time, dynamic force $-F_d$ should appear on the Earth as there is no external force to balance $-F_s$. A mass in free fall in a gravitational field created by other masses will always produce reaction forces on those masses equal to its weight. We see that the principle of action and reaction applies to the free fall, and masses accelerate relative to the centre of masses.

When we consider the acceleration of the mass on the Earth, we traditionally consider only force that acts on the accelerating mass and response of that mass. We do not follow the other end of the chain of forces that ends up on the Earth. Nothing is holding the Earth immovable in space. The force acting on the Earth will accelerate the Earth. The acceleration of the Earth under the influence of a small force will be negligible, but it is present. When we consider all forces in accelerating the mass on the Earth, we see that the inertial forces are balanced and all masses accelerate relative to the centre of all masses. The principle of the action and reaction applies to the inertial forces in every acceleration, regardless of whether it is a free fall or not. Does the acceleration of the mass in space cause inertial forces, and we need to mathematically account for the balance of forces by considering all the masses involved? Or is the balance of forces the consequence of some physical interaction between the moving masses that ensures the balance of forces? The hypothesis of the dynamic gravitational field explores the possibility that the physical interaction causes inertial force and suggests the experiment to test that hypothesis in the laboratory.

Under free fall conditions on Earth, there is no external force on the falling body. The forces on the falling mass should balance internally. According to the hypothesis of the dynamic gravitational field, the weight of the mass created by the static gravitational field should be balanced by the inertial force created by the dynamic gravitational field. If we assign the mass of the Earth to m_2 in equation (14a), we see that the acceleration of any mass will be equal to the strength of the static gravitational field for the Earth. The centre of masses (for the Earth and the falling body) will not change due to the mass of the falling body. The movement of the Earth will be negligible as the falling body has negligible mass compared to the Earth's mass. For us, any mass will fall to the Earth with the same acceleration even when the masses falling to the Earth are very different.

When reviewing Mach's principle [Reinhardt 1973], one of the conclusions was that Mach's principle is incompatible with celestial mechanics if nearby masses dominate the determination of the inertial mass. This conclusion is correct if we apply Mach's principle to the mass's inertial property rather than considering the influence of nearby masses on the inertial force. What we observe in the experiments are inertial forces. Assigning inertial property to the mass is Galileo's and Newton's explanation of the nature of those forces. This article shows that nearby masses can play a role in creating the inertial force, and Mach's principle can be compatible with celestial mechanics when applied to the inertial force rather than to the inertial property of the mass. When inertial forces are explained by the force field, the inertial property of the mass plays the same role in gravitation as an inductance in electromagnetic theory.

5 Design of the experiment to detect the dynamic gravitational field

The magnitude of the predicted dynamic gravitational field is large enough to be detected by a suitably designed laboratory experiment. If we want to measure the dynamic gravitational field for a ball rolling down an incline or for a ball dropped from some height in a laboratory, the field's magnitude and the event's duration would be too small to be detected by a simple instrument. If we use a spinning gyroscope, as shown in Fig. 5, we will have a mass dm accelerating in the same location relative to us all the time. This mass would create a dynamic gravitational field as per equation (12). The integral of the dynamic gravitational field for all the masses in the gyroscope would not change with time. It depends only on the position in space. An instrument like a torsion balance could detect the field. The instrument should be able to work in an environment where there are disturbances caused by vibrations of the gyroscope, as it can never be fully balanced.

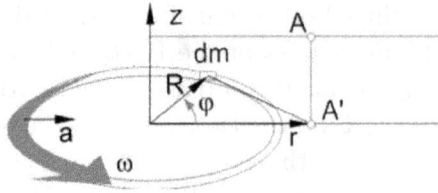

Fig. 5 Gyroscope for the generation of the dynamic gravitational field

The strength of the dynamic gravitational field around the rotating gyroscope shown in Fig. 5 can be found by integrating contributions of all masses dm along the perimeter of the gyroscope. Using the cylindrical coordinates, the expression for the strength of the dynamic gravitational field that the gyroscope creates in point A could be written as

$$\vec{g}_d = \int_{-\pi}^{\pi} G \frac{\omega^2 R}{14.72} \frac{\cos(\varphi)\lambda R d\varphi}{(z^2 + r^2 + R^2 - 2rR\cos(\varphi))} \tag{15}$$

The dynamic gravitational field for a gyroscope, which has a mass of 10 kg uniformly distributed along the perimeter of the circle with a radius of 0.3 m, like a bicycle wheel, is shown in Fig 6. The magnitude of the dynamic gravitational field is calculated using equation (15).

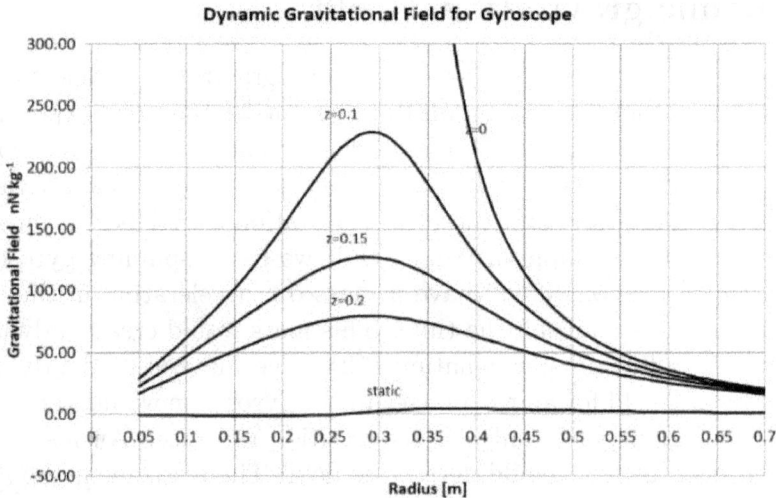

Fig. 6 Strength of the dynamic gravitational field around the gyroscope

The gyroscope spins at six revolutions per second or 360 rpm. The wheel's perimeter is moving at about $40 \, \text{kmh}^{-1}$ and has an acceleration of about 426 ms^{-2}. On the y-axis in Fig. 6, we have the field strength in nNkg^{-1}, and on the x-axis, we have the distance from the centre of the gyroscope in meters. The positive value of the field is in the direction toward the centre of the gyroscope. The distance on the z-axis in meters is marked for every graph as a parameter. The component of the static gravitational field pushing towards the centre of the gyroscope at $\Delta z = 0.1$ m is also shown for comparison. The strength of the dynamic gravitational field for the gyroscope is comparable to the strength of the static gravitational field in the Cavendish experiment, which was about 200 nNkg^{-1}.

The instrument should be able to detect the gravitational field above the noise level. We need to allow for casing around the instrument probe that will protect the probe from the air movement and electrostatic effects so that we cannot come too close to the probe with the gyroscope.

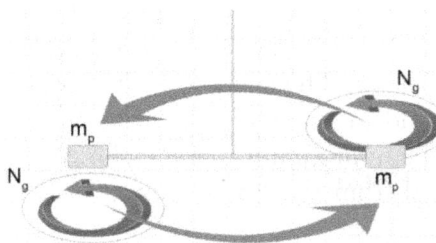

Fig. 7 Modified Cavendish experiment

The experimental setup that can detect the dynamic gravitational field is shown in Fig. 7. It consists of a torsion balance scale with the masses of the probes m_p and two gyroscopes. The gyroscope should rotate at the operating speed N_g. The magnitude of the field should be measured by moving the gyroscopes around the probes, similarly to the Cavendish experiment, so that the gravitational field acting on the probes changes direction [Cavendish 1798] and [Chen and Cook 2005]. There could be other experiments that could detect the dynamic gravitational field.

6 Properties of space around the moving masses

The hypothesis of the dynamic gravitational field removes space as an active participant in creating the inertial force. The properties of the space are not postulated but can be determined and experimentally explored. The observations have shown that the propagation of light, or electromagnetic waves, is influenced by the distribution of masses. We know that light rays bend around the sun due to the presence of the sun's gravitation. According to the hypothesis of the dynamic gravitational field, the effects of the mass acceleration are not confined to the interior of the mass; they are felt in the space around the accelerating mass. The dynamic gravitational field, if it exists, should also influence light propagation. To find out how, let us integrate equation (9) with time

$$\vec{v}_{d1} = G\frac{m_1}{r^2} \frac{\vec{v}_1 + C}{\int_V G\left(\rho_i/r_{1i}^2\right) dV} = G\frac{m_1}{r^2} \frac{\vec{v}_1 + C}{\sum G\left(m_i/r_{1i}^2\right)} \tag{16}$$

In this equation, we have velocities instead of accelerations. The velocity v_{dl} in this equation is the velocity of the space caused by the velocity v_1 of the mass m_1. The denominator in the equation defines the influence that the other masses have on the velocity of space caused by the movement of mass m_1.

The constant of integration C could be selected to measure the movement of space observed from one mass. If we observe stars from the Earth, declaring Earth's velocity zero would be the closest to how we observe the universe. The space velocity caused by the movement of other masses could be calculated relative to the Earth. Assuming that the Earth and sun are point masses, shown in Fig. 8, the velocity of space as observed from the Earth caused by the motion of the sun is given by

$$\vec{v}_{sp} = G\frac{m_s}{d^2 + r^2} \frac{\vec{v}_s}{\left(G\left(m_e/r^2\right) + G\left(m_s/\left(d^2 + r^2\right)\right)\right)}, \tag{17}$$

where is

- v_{sp} – velocity of the space observed from the Earth $[\mathrm{m\,s^{-1}}]$
- G – gravitational constant $[6.67*10^{-11}]$
- m_s – the mass of the sun $[1.99*10^{30}\ \mathrm{kg}]$
- m_e – the mass of the Earth $[5.97*1024\ \mathrm{kg}]$
- d – the distance between the sun and the Earth $[1.5*10^{11}\ \mathrm{m}]$
- r – distance from the centre of the Earth $[\mathrm{m}]$
- v_s – velocity of the sun relative to the Earth $[30\ \mathrm{km\,s^{-1}}]$

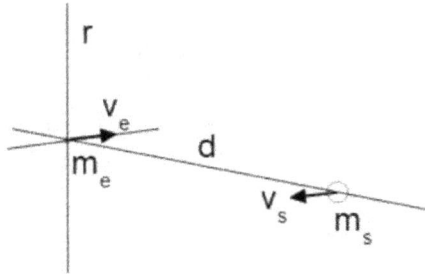

Fig. 8 Diagram of the velocities in the Earth/sun system

In the Earth/sun system, we can consider the velocity of the space on the line which goes through the Earth's centre and is perpendicular to the plane in which the Earth rotates around the sun. The velocity of the space in the Earth-sun system would be determined by equation (17). The velocity of the space on the surface of the Earth, along the vertical line in Fig. 8, caused by the sun's movement, would be approximately 12 ms^{-1}. The velocity of the space far from the Earth would asymptotically approach the sun's velocity as the denominator in the equation for the speed of the space changes. The velocity of the space around the Earth is shown in Fig. 9.

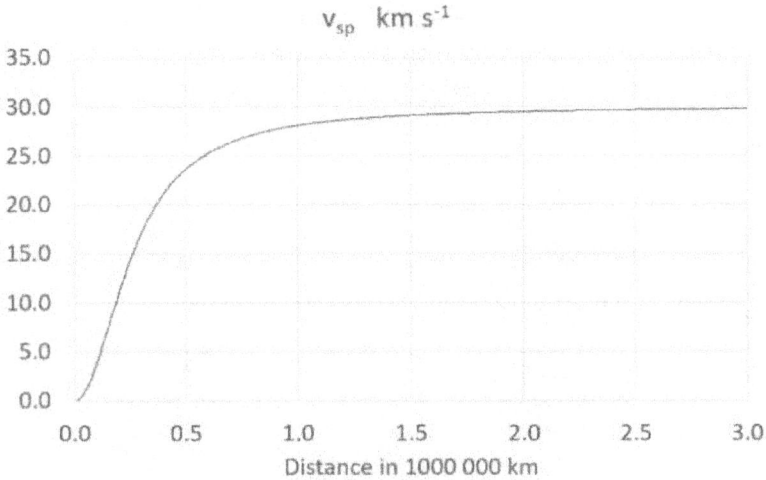

Fig. 9 Velocity of the space in the Earth/sun system relative to the Earth

The above consideration has linked the presence and movement of the mass to the properties of space. We could see that the velocity of the space varies with the distance. At about three million kilometres or 0.02AU from the Earth, the space is fixed relative to the sun. Using the hypothesis that accelerating mass develops a force field around it points to the existence of an inhomogeneous physical space or ether dragged by the moving masses. The mass distribution and the acceleration of the masses cause the curvature of space. The equation that describes the properties of space with moving mass and the static and dynamic gravitational field is

$$\vec{v}_{d1} = G \frac{m_1}{r^2} \frac{\vec{v}_1 + (\vec{g}_{s1} + \vec{g}_{d1}) \, dt + C}{\int_V G \left(\rho_i / r_{1i}^2\right) dV} \tag{18}$$

The movement of the mass or the gravitational field that is perpendicular to the velocity of light will cause a change in the direction of the light ray. The movement of the mass or the gravitational field parallel to the velocity of the light ray will cause a doppler effect.

James Bradley observed the movement of the Earth in space in nature in 1727. Bradley noticed that stars in the zenith appear to move for about $9.6*10^{-5}$ radians in the direction of motion of the Earth around the sun. The effect was named aberration of the light.

The propagation of light in space with a moving medium was a big question in nineteenth and beginning of twentieth-century physics. The belief was that there exists the static ether that carries the light. Because of the observed aberration of light, it was believed that the Earth travels in the ether at $30 \, \mathrm{kms}^{-1}$. Physicists wanted to detect the movement of the Earth in the ether by an experiment involving light. Several experiments were carried out to determine that movement. I will mention the following:

- Hoek experiment

- Fizeau experiment

- Michelson-Morley experiment

- Sagnac experiment

- Michelson-Gale experiment

The experiments failed to confirm that the Earth is moving through space at $30 \, \mathrm{kms}^{-1}$. The results of these experiments and observations agree with the assumption that Einstein has made that the ether doesn't exist and the speed of light is constant in every inertial frame of reference. They are also consistent with the assumption that there exists a local ether that carries the light. The movement of the Earth entrains the local ether. In [Su 2001], Su has shown that the results of the light experiments and corrections of GPS signals can be explained by considering a local ether frame that is stationary to the Earth or the sun. He calls them Earth cantered inertial frame ECI or heliocentric reference frame. He uses such reference frames to explain the propagation of radio waves and GPS corrections using the classical ether concept. Su did not offer a way of calculating the speed of ether. The equation (16) could present a way to calculate the speed of ether and determine the frame of reference from mass distribution and movement of the masses relative to each other. The propagation of light in local ether has an analogy with the movement of the air inside the flying aeroplane. That air will carry sound waves with it, so the speed of sound relative to a plane is not affected by the aircraft's speed. We can't detect mass movement by measuring the speed of light in the local ether as the moving mass carries the ether.

Gravitational waves have been detected recently [Abbott et al. 2016]. According to the dynamic gravitational field hypothesis, the gravitational wave is a movement of the local ether. The rapid movement of the masses in the universe causes movement of the local ether, which carries the light. The dynamic gravitational field hypothesis does not address how gravitation propagates. It just explains the nature of gravitational waves.

7   Dark matter or dynamic gravitational field

It was observed that the rotational speed of the stars in the galaxy is much higher than what would be expected from the baryonic mass when we calculate it using Newton's method. Fig. 10, reproduced from [Karukes et al. 2015], shows the velocities of the stars in the spiral galaxy NGC 3198 as a function of the distance from the galaxy's centre.

The black dots with the bars show measured velocities with measurement tolerance ranges. The line that decays with the distance shows velocities as predicted by Newton's method. The predicted velocities decay with the distance from the galaxy's centre while measured velocities remain high. The measured velocities of the stars point to the additional gravitational pull towards the galaxy's centre. The hypothesis of dark matter, an additional galaxy mass that does not interact with light, is introduced to explain the discrepancy in measured and predicted velocities of stars.

Milgrom [Milgrom 1983] has proposed a modification of the second Newton's axiom that explains the observed velocities of stars without

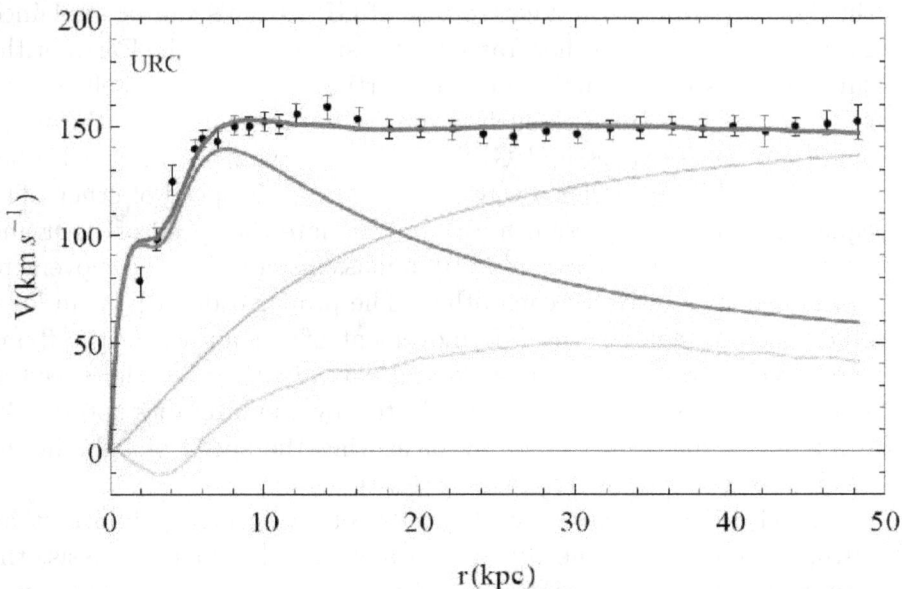

Fig. 10 Velocity of the stars in the galaxy

the need for dark matter. He has modified it with a function that changes inertial mass with acceleration. This modification abandons the principle of equivalence of gravitational and inertial mass, which is the foundation of the general theory of relativity. For the galactic scale, for which acceleration is small, Milgrom has reduced the inertial mass in the second Newton's axiom proportionally to the reduction in acceleration. Milgrom did not provide a physical reason for the decrease in inertial mass. If proven by the experiment, the dynamic gravitational field could explain why the inertial mass appears smaller.

Analysis of the velocity of stars for 240 galaxies [Lelli et al. 2017] has found that the velocities follow the same radial acceleration relation, which deviates from that expected by Newton's method. One of the implications of the analysis for the alternative theories of gravitation is that we may need a new fundamental law of physics to explain observations.

In this article, I have removed the second Newton's axiom altogether and replaced it with the hypothetical force field. According to the dynamic gravitational field hypothesis, the inertial property of the mass that Galileo introduced and Newton used in his second axiom is analogous to the property of inductance in the electromagnetic theory. There is a field that produces the effects of inertia.

The dynamic gravitational field provides an alternative explana-

148

tion of the discrepancy in measured and predicted velocities of stars that don't require the dark matter hypothesis. The galaxy can be treated as a giant gyroscope that produces a dynamic gravitational field. Detection of additional gravitational pull can be interpreted as observational evidence supporting the claim that a rotating gyroscope develops a gravitational field acting toward its centre. The dynamic gravitational field will explain additional gravitational pull if the experiment confirms its existence.

8 Conclusion

This article has presented the theory of dynamic gravitational field based on the hypothesis that every accelerating mass creates a force field around it. The inertial force is explained as a consequence of this dynamic gravitational field. The equation for the gravitational field created around an accelerating mass is derived using the principle of action and reaction. The equation assumes that the law of motion $F = ma$ is valid everywhere in the universe. It has been shown that properties of the inertial force on Earth calculated using the dynamic gravitational field agree with the observations in nature. Also, it has been shown that the movement of the planets in the solar system calculated using the equations for the dynamic gravitational field agrees with the observations. On the galactic scale, the dynamic gravitational field explains the observed velocities of stars without the need for the dark matter hypothesis. The magnitude of the dynamic gravitational field that produces inertial force is large enough to be detected in the laboratory. The experiment that can prove or disprove the existence of the dynamic gravitational field is described.

The fact that we do not have experimental evidence that the dynamic gravitational field exists is not proof that such a field does not exist. The claim that the principle of action and reaction does not apply to the inertial force is a big claim that is not supported by experimental evidence. We need to perform an experiment that will show the nature of the inertial force.

The dynamic gravitational field is quasistationary. The theory of the dynamic gravitational field is not a complete gravitational theory. The question of how the gravitational field propagates in space is not considered here. That will be an open question if a dynamic gravitational field is detected. There is a saying that if something in fundamental physics could be tested by the experiment, it should be tested. We shouldn't rely on logic and theoretical consideration to answer questions about nature. I would urge experimental physicists to perform the experiment suggested and resolve the criticism of the concept of inertia in the laboratory. The experiment may resolve the issue of dark matter that physicists currently investigate.

The article describes the laboratory experiment that can differentiate between two concepts for creating inertial force. The hypothesis that mass possesses inertial properties is an old and well-developed hypothesis. The hypothesis that the force field causes inertial force is new, and the article develops it. We could argue in favour of one or another hypothesis, but that would not change the way nature acts. The point worth discussing is whether we need to experiment or not. The experiment suggested here has scientific merit and will contribute to science regardless of the experiment's outcome. If the experiment shows that the principle of action and reaction does not apply to the inertial force, it will support the case for the general theory of relativity. We will confirm that reaction to the inertial force does not exist by showing that the opposite hypothesis is false. In mathematics, such proof is referred to as proof by contradiction. This result will significantly contribute to the general theory of relativity as the lack of reaction to the inertial force is a serious criticism that even Einstein acknowledges. If the experiment detects a dynamic gravitational field, it will open a new avenue for gravitational research. We will know more about the inertia once we do the experiment, whatever the result.

References:

Abbott B. P. et al., The Observation of Gravitational Waves from a Binary Black Hole Merger, *PRL* 116, 061102 (2016) DOI:10.1103/PhysRevLett.116.061102

Cavendish H., "Experiments to Determine the Density of the Earth" http://dx.doi.org/10.1017/CBO9780511722424.019

Chen Y. T., Cook A., *Gravitational Experiments in the Laboratory*, Cambridge University Press, 2005 http://dx.doi.org/10.1017/CBO9780511563966

Einstein A., *The Meaning of Relativity* ISBN 978-1-78139-864-7 Benediction Classics, Oxford 2017

Everitt C W F et all, The Gravity Probe B test of general relativity, *Class. Quantum Gravity.* 32 (2015) 224001 (29pp) https://doi.org/10.1088/0264-9381/32/22/224001

Karukes E. V. et all, The dark matter distribution in the spiral NGC 3198 out to 0.22 R_{vir} − A& A578, A13 (Apr 2015) https://doi.org/10.48550/arXiv.1503.04049

Lelli F., McGaugh S. S., Schombert J. M., Pawlowski M. S., One Law to Rule Them All: The Radial Acceleration Relation of Galaxies, *The Astrophysical Journal*, 836:152 (23pp), 2017 February 20 https://doi.org/ 10.3847/1538-4357/836/2/152

Mach E., *The Science of Mechanics* ISBN 0-87548-202-3 Open Court Clasics 1989

Milgrom M., A Modification of the Newtonian Dynamics as a Possible Alternative to the Hidden Mass Hypothesis, *The Astrophysical Journal*, 270:365-370, 1983 July 15 doi:10.1086/161130

Reinhardt M. Berichte, Mach's Principle - A Critical Review, *Zeitschrift für Naturforschung* A, vol. 28, no. 3-4, 1973, pp. 529-537. https://doi.org/10.1515/zna-1973-3-431

Ruggiero M. L., Tartaglia A.: Test of gravitomagnetism with satellites around the Earth, *The European Physical Journal Plus*, volume 134, Article number: 205 (2019) https://doi.org/10.1140/epjp/i2019-12602-6

Sciama D. W., On The Origin Of Inertia, *Monthly Notices of the Royal Astronomical Society*, Volume 113, Issue 1, February 1953 https://doi.org/ 10.1093/mnras/113.1.34

Su C. C., A local-ether model of propagation of electromagnetic wave, Sep. 2001 *Eur. Phys. J. C* 21, 701-715 (2001) https://doi.org/10.1007/s100520100759

8 PROTOGRAVITY: A QUANTUM-THEORETIC PRECURSOR TO GRAVITY

DANIEL SHANAHAN

Abstract   As a consequence of the twin effect – the slower aging of the twin who makes a return trip compared with that of her brother who stays at home – an orbiting mass has a reduced proper time and thus a binding energy. Referring to this binding effect as *protogravity*, I argue that it was sufficient in itself to explain the preference of matter for bound rather than free motion in the early universe. I then show, from a consideration of the constraints imposed by laws of conservation and the principle of relativity, that this protogravity is also able to explain the important Schwarzschild metric, and thus the effects of gravity so far as these can be determined with accuracy from the Earth. I argue therefore that gravity need not involve a constraining geometry or any other extraneous force or effect, but is adequately explained as an emergent consequence of the manner in which elementary particles adapt to a change of inertial frame. Because the twin effect may be explained from the evolution of phase described by the de Broglie wave, this interpretation of gravity would provide gravity and quantum mechanics with a common origin in the wave-like nature of the elementary particles.

Keywords: emergent gravity, the twin effect, Schwarzschild metric, relativity of simultaneity, de Broglie wave, principle of relativity

> *It is said that more than 200 theories of gravity have been put forward ...*
> Sir Arthur Eddington, writing in 1920 [1], p. 64.

1   Introduction

I explore the possibility that gravity has its origin with quantum mechanics in the manner in which the wave structure of a elementary particle must adapt to a change of inertial frame of reference.

A. S. Stefanov, G. Dupuis-Mc Donald (Eds), *Spacetime Conference - 2022*. *Selected peer-reviewed papers presented at the Sixth International Conference on the Nature and Ontology of Spacetime, 12 - 15 September 2022, Albena, Bulgaria* (Minkowski Institute Press, Montreal 2023). ISBN 978-1-989970-96-6 (softcover), ISBN 978-1-989970-97-3 (ebook).

A particle experiences a range of changes as it changes inertial frame. For a massive particle, these are the changes in length, time and simultaneity described by the Lorentz transformation and, at what will be regarded here as the more fundamental level, the corresponding variations in wavelength, frequency and phase defined for quantum mechanics by the de Broglie wave.

Whether deduced from the Lorentz transformation or from the de Broglie wave, these changes are the source of what has been called "the twin effect" – the slower aging of the twin who makes a return trip compared with that of her brother who stays at home. To a observer who is stationary with respect to an orbiting particle and is also beyond the reach of gravity (the notional observer at infinity), the particle has, from this dilation of time[1], a reduced frequency ω_E (its Einstein frequency) and, from the Planck-Einstein relation,

$$E = \hbar\omega_E, \tag{1}$$

a correspondingly reduced energy E (where $\hbar$ is Planck's constant).

This loss of proper time is empirically well-established. It was demonstrated by the Earth-circling clocks of the Hafele-Keating experiment [2] and is observed in the enhanced half lives of cosmic rays and accelerated muons [3]. It is evidenced every minute of every day, and to a high degree of accuracy, by the atomic clocks of global positioning systems[2].

It follows from the twin effect that an orbiting particle, and consequently any orbiting object, has a binding energy, and it is this binding effect that I have referred to above as *protogravity*. If there were no other form of gravitational attraction in the universe, a system of mutually orbiting objects would have, from their orbital motion alone, a binding energy sufficient to hold those objects to their paths.

As thus described, this protogravity is not yet the gravity that is actually experienced. A stationary object also feels the effects of gravity, while a moving object experiences a dilation of time, not only from its movement relative to a gravitating mass, but from its proximity to that mass. Nor is the binding force due to the twin effect usually thought of as a form of gravity. For instance, in discussions of global

---

[1]Taking a cosmic stance, time is lost. But taking the view that wih less time expended, there is more remaining (for life or half-life), the commoner usage is that time becomes dilated (expands).

[2]There are over a hundred GPS satellites orbiting the Earth. The United States, Russia, China and the European Union maintain international systems, while Japan and India have local systems

positioning systems, a distinction is drawn between the dilation of time due to the orbital motion of the satellite, which is taken to be a consequence of special relativity, and the dilation due to the depth of the orbit within the gravitational influence of the Earth, as for example in Ashby [4].

Yet this protogravity contributes a binding force that would have been sufficient in itself to explain the tendency of matter to favour bound rather than unbound motion in the early universe. And as I will show, from a consideration of the further constraints imposed by the principle of relativity and the conservation of energy and angular momentum, this binding force is able to explain the important Schwarzschild metric, and in so doing, the effects of gravity so far as these can be ascertained with reasonable certainty from the Earth.

The twin effect will provide the underlying mechanism of gravitational attraction, while the principle of relativity and laws of conservation will determine the relative strength of this mechanism from one situation to another.

The dilation of time experienced by an object that is stationary with respect to a gravitating mass will be explained by the loss of momentum and thus energy that such an object experiences when it is brought to rest after falling from infinity, a loss which in accordance with the Planck-Einstein relation (Eqn. (1) above) is accompanied by a loss of frequency and thus of time. The further dilation experienced by an object that is moving within the influence of gravity will then follow from the stipulation, pursuant to the principle of relativity, that moving and stationary particle have the same interactions and dynamic relationships within that influence as they do when beyond it.

Why should this protogravity be deemed, as advertised above, a *quantum-theoretic* precursor of actual gravity? When the twin effect is explained (see Sect. 2 below) from the loss of phase in the direction of travel defined by the de Broglie wave, it acquires a common origin with quantum mechanics. It was de Broglie's prediction of this "matter wave" in his famous thesis of 1923 that allowed a quantum mechanics in which massive particles are treated in terms of evolving wave characteristics. The Schrödinger equation and other equations of quantum mechanics for massive particles, including the Dirac and Pauli equations, were originally conceived as equations for the de Broglie wave, see Bloch [5] and Dirac [6]. Were it not for the de Broglie wave, there would not be a quantum mechanics, not at least a quantum mechanics for massive particles.

Even so, the explanation of gravity that I will offer in this paper could be presented with no mention at all of quantum theory or wave mechanics beyond the reference to the Planck-Einstein relation already made above. And this is the way in which I will initially present the argument. But this interpretation of gravity was initially conceived from a consideration of wave structure. Moreover, it will become apparent as the argument proceeds that in this interpretation, gravity is not strictly speaking a fundamental effect, but emergent from other laws of Nature, namely the Lorentz transformation, laws of conservation, the principle of relativity and the Planck-Einstein relation. These laws and principles are well-established, but are ultimately empirical and thus brute and unexplained, as also, I suggest, is the mysteriously superluminal de Broglie wave. In the concluding sections of the paper, I will endeavour to show that some at least of this miscellaneous collection of laws and phenomena have a common origin with gravity and quantum mechanics in the underlying wave structure of matter and radiation.

In this paper, I will not take gravity beyond the Schwarzschild metric. It seems unlikely that this "bootstrap" approach to gravity, in which orbital motion is self sustaining, would replicate general relativity in all possible situations. And that perhaps is reason in itself for pursuing this proposal. It is in relation to orbital motion, notably in its failure to explain galactic rotation curves and the origin of the angular momenta of the galaxies, that Einstein's theory seems to require further investigation.

The twin effect is counterintuitive and as a preliminary step I will discuss in the next section how this dilation of time is related to the relativity of simultaneity.

2   The twin effect

The twin effect will be derived in three ways, these being in the historical order of their origination: (a) from the Lorentz transformation; (b) in Minkowski spacetime; and (c) as a consequence of the dephasing described by the de Broglie wave.

What is counterintuitive here is that an object should have an energy that is less when it is moving than when it is stationary. And of course, to an observer that a massive particle is passing, that particle does have an increased frequency and correspondingly increased energy,

$$E = \gamma E_o = \gamma \hbar \omega_o, \tag{2}$$

where E_o and ω_o are, respectively, the energy and frequency of the particle in its rest frame, while γ (in units in which $c = 1$) is the Lorentz factor,

$$(1 - v^2)^{-\frac{1}{2}}, \tag{3}$$

and it is this enhanced energy and its associated momentum that a moving object would bring to a collision with that observer.

But as an orbiting particle continues on its way, it is also experiencing the failure of simultaneity described by the Lorentz transformation, and this accumulates as a slowing of time and consequent decrease in energy per orbit that has (approximately at non-relativistic velocities) twice the magnitude of the increase in energy described by Eqn. (2).

(a) from the Lorentz transformation: These competing effects are combined in the time component,

$$dt' = \gamma \left(dt - v ds \right),$$

of the Lorentz transformation, which by the substitution,

$$ds = v dt,$$

describes a reduced proper time,

$$d\tau = dt' = (1 - v^2)^{\frac{1}{2}} dt,$$

giving for a complete orbit,

$$\tau = \oint (1 - v^2)^{\frac{1}{2}} dt \tag{4}$$

and thus, for an orbiting particle of rest mass m_o, a reduced energy,

$$m_o P^{-1} \oint (1 - v^2)^{\frac{1}{2}} dt,$$

where P is the orbital period as considered by the notional observer at infinity.

For the purpose of comparison, it will be more convenient to use from this point the energy per unit mass (referred to in astrodynamics as the *specific mechanical energy* ε, see *Bate et al* [7], at p. 15), which for the twin effect will be designated,

$$\varepsilon_{twin} = P^{-1} \oint (1 - v^2)^{\frac{1}{2}} dt,$$

where the binding energy (or mass defect), which is expressed in the negative, is,

$$\Delta \varepsilon_{twin} = P^{-1} \oint (1 - v^2)^{\frac{1}{2}} dt - 1. \tag{5}$$

(b) in Minkowski spacetime: In a spacetime diagram (Minkowski [8]), the proper time τ of a test particle following a time-like trajectory may be written,

$$\tau = \int \left(-ds^2\right)^{\frac{1}{2}} = \int \left(g_{uv}\, dx^u dx^v\right)^{\frac{1}{2}}, \tag{6}$$

where g_{uv} is the relevant metric, and u and v signify *four*-coordinates.

For the Minkowski metric, g_{uv} is the metric tensor,

$$\eta_{uv} = \begin{vmatrix} -1 & 0 & 0 & 0 \\ 0 & 1 & 0 & 0 \\ 0 & 0 & 1 & 0 \\ 0 & 0 & 0 & 1 \end{vmatrix}$$

so that Eqn. (6) becomes, in differential form, the invariant interval,

$$-d\tau^2 = ds^2 = -dt^2 + dx^2 + dy^2 + dz^2. \tag{7}$$

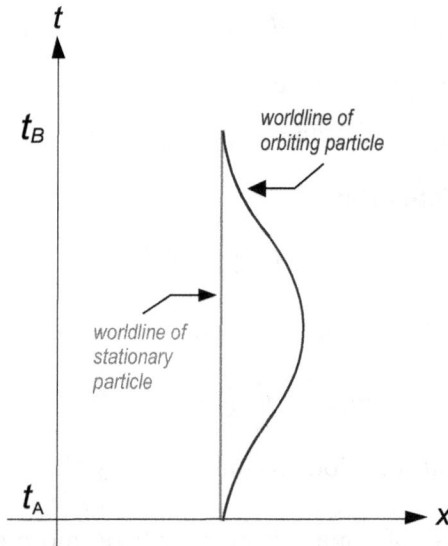

Figure 1: The proper time experienced by an orbiting particle is less than that for a stationary particle.

In the spacetime diagram of Fig. 1, a particle follows a curved (and thus accelerated) path between events A and B. Because proper time is a scalar invariant and thus has a magnitude on which all observers must agree, the axes of the diagram can be chosen (as they have been in the drawing) so that A and B lie conveniently on the t axis, which is thus the world line of an observer who in this frame of reference is stationary at $x = 0$.

In its own co-moving inertial frame, the particle is stationary. Thus, for any infinitesimal interval ds along its worldline,

$$dx = dy = dz = 0,$$

and Eqn. (7) reduces to,

$$d\tau = dt.$$

But to the observer at $x = 0$, for whom the particle is moving at the velocity $v(t)$, Eqn. (7) becomes, after division by dt,

$$\frac{d\tau}{dt} = \left(1 - v^2\right)^{\frac{1}{2}}, \tag{8}$$

whereupon for the complete worldline of the orbit (along the curved path between A and B), we again have Eqn. (4), that is,

$$\tau = \oint \left(1 - v^2\right)^{\frac{1}{2}} dt$$

as was deduced from the Lorentz transformation, and which as we have seen leads to the binding energy $\Delta\varepsilon_{twin}$ of Eqn. (5).

(c) from the de Broglie wave: A massive particle has from the Planck-Einstein relation (Eqn. (1) above), an associated frequency (the Einstein frequency ω_E), and from the de Broglie relation,

$$p = \hbar\kappa_{dB}, \tag{9}$$

a wave number κ_{dB} (the de Broglie wave number), where E and p are, respectively, the energy and the momentum of the moving particle.

Frequency ω_E and wave number κ_{dB} define for the moving particle, its de Broglie wave,

$$\psi_{dB} = e^{i(\omega_E\, t - \kappa_{dB}\cdot\mathbf{r})},$$

from which the evolution of phase per orbit is,

$$\varphi_{orbit} = \oint [\omega_E\, dt - \kappa_{dB}\, ds],$$

which on making the substitution,

$$ds = v\, dt,$$

and using relations (1) and (9) becomes,

$$\varphi_{orbit} = \omega_0 \oint (1 - v^2)^{\frac{1}{2}}\, dt,$$

so that the energy per unit mass is,

$$\varepsilon_{twin} = \frac{\omega_0 \oint (1 - v^2)^{\frac{1}{2}} \, dt}{\omega_0 P}$$

$$= P^{-1} \oint (1 - v^2)^{\frac{1}{2}} \, dt,$$

giving for the binding energy per unit mass,

$$\Delta\varepsilon_{twin} = P^{-1} \oint (1 - v^2)^{\frac{1}{2}} \, dt - 1,$$

which is the result obtained above as Eqn. (5) from the Lorentz transformation, but derived now from the evolving wave characteristics of the particle.

Thus all three derivations of the twin effect lead to the same dilation of time and binding energy $\Delta\varepsilon_{twin}$.

3   Newtonian gravity

As a further step toward the Schwarzschild metric, I will establish in this section that for the closed elliptical orbits of Newtonian gravity, binding energies derived from the twin effect correspond exactly with those deduced from the centripetal force supposed by Newton's universal law of gravitation.

From observations recorded by Tycho Brahe, Kepler had deduced that the planets move in accordance with three laws:

  1. The orbit of a planet is an ellipse with the sun at one focus;

  2. A line drawn from a planet to the sun sweeps out equal areas in equal times; and

  3. The square of the period of a planet is proportional to the cube of its mean distance from the sun.

Newton then showed in the *Principia* [9] that if angular momentum is conserved in accordance with the second of Kepler's laws (the law of areas), the elliptical paths described by the first of those laws are consistent with the existence of a centripetal force acting directly between massive objects and having a strength varying inversely with the square of the distance between those masses. This is Newton's universal law of gravitation,

$$F = G\frac{m_1 m_2}{r^2},$$

where G is the universal gravitational constant, m_1 and m_2 are the masses, and r is the distance between them[3].

In the *Principia*, Newton offered no explanation for this attractive force. He famously declared "hypotheses non fingo", as in the recent translation by Cohen and Whitman [9], at p. 276:

> I have not as yet been able to deduce from phenomena the reason for these properties of gravity, and I do not feign hypotheses. For whatever is not deduced from the phenomena must be called a hypothesis; and hypotheses, whether metaphysical or physical, or based on occult qualities, or mechanical, have no place in experimental philosophy. In this experimental philosophy, propositions are deduced from the phenomena and are made general by induction. The impenetrability, mobility, and impetus of bodies, and the laws of motion and the law of gravity have been found by this method. And it is enough that gravity really exists and acts according to the laws that we have set forth and is sufficient to explain all the motions of the heavenly bodies and of our sea.

In other writings, Newton described the notion that a force could act at a distance "without the mediation of any thing else" as "so great an absurdity that .... no man who has in philosophical matters a competent faculty of thinking can ever fall into it" [10]. And outside the *Principia*, he did consider possible explanations for this force, though seems ultimately to have seen in the orbits of the planets, the guiding hand of a divine providence.

It is a simple matter to show that the binding energies predicted by Newton's central force correspond exactly with those obtained from the twin effect, that is,

$$\Delta\varepsilon_{Newton} = \Delta\varepsilon_{twin}.$$

I will begin with a circular orbit, and then consider the more general case of an elliptical orbit.

From Eqn. (5) of the preceding section, the binding energy per unit mass due to the twin effect is,

---

[3]To glimpse the enormity of Newton's achievement, it is necessary to visit the Principia, and to understand that to pursue his many proofs and theorems, he had first to invent the calculus as well as the notion of mass as a measure of substance [9], Def. 1. And all this at a time when it was still possible for the rival theory of Descartes to explain the motions of the planets by circulating fluxes of a mysterious fluid.

$$\Delta\varepsilon_{twin} = P^{-1} \oint (1 - v^2)^{\frac{1}{2}} \, dt - 1,$$

which in the Newtonian approximation where $v << c$ becomes,

$$\Delta\varepsilon_{twin} = -P^{-1} \oint \frac{1}{2}v^2 dt. \tag{10}$$

For a circular orbit, v is constant, so that,

$$\Delta\varepsilon_{twin} = -\frac{1}{2}v^2. \tag{11}$$

Turning now to Newtonian gravity, the binding energy per unit mass, $\Delta\varepsilon_{Newton}$, is the sum of the object's kinetic and potential energies, the latter being taken to be zero at infinity. Thus,

$$\Delta\varepsilon_{Newton} = T + V = \frac{1}{2}v^2 - \frac{GM}{h}, \tag{12}$$

where h is the distance of the unit mass from the centre of the central mass, (r being reserved here for the coordinate distance, as in the Schwarzschild metric, that is to say the distance observed by the notional observer at infinity).

For a circular orbit (*Bate et al* [7], at p, 34)

$$v = \left(\frac{GM}{h}\right)^{\frac{1}{2}},$$

and Eqn. (12) becomes,

$$\Delta\varepsilon_{Newton} = \frac{GM}{2h} - \frac{GM}{h} = -\frac{GM}{2h}, \tag{13}$$

and thus for a circular orbit it follows from Eqns. (11) and (13) that as required,

$$\Delta\varepsilon_{Newton} = \Delta\varepsilon_{twin} = -\frac{1}{2}v^2 = -\frac{GM}{2h}.$$

For an elliptical orbit (see, for instance, Logsdon [11], at p. 30), the velocity is given by Newton's *vis-viva* formula,

$$v = \left[GM\left(\frac{2}{h} - \frac{1}{a}\right)\right]^{\frac{1}{2}}, \tag{14}$$

162

where a is the semi-major axis of the ellipse.

From Eqns. (12) and (14)

$$
\begin{aligned}
\Delta \varepsilon_{Newton} &= \frac{1}{2}v^2 - \frac{GM}{h}, \\
&= \frac{GM}{h} - \frac{GM}{2a} - \frac{GM}{h}, \\
&= -\frac{GM}{2a}.
\end{aligned}
$$

Thus the binding energy is independent of the eccentricity e of the ellipse, depending only for a central mass M on the magnitude of the semi-major axis a, and this is so for the limiting case of a straight line orbit (where the maxima are at $h = \pm 2a$), and that of a circular orbit (where $h = a$).

For an elliptical orbit, we have from Eqns. (10) and (14),

$$
\Delta \varepsilon_{twin} = P^{-1} \oint [\frac{GM}{h} - \frac{GM}{2a}]dt,
$$

but it follows from the virial theorem (see Goldstein [12], at p. 85) that for orbits consistent with an inverse square law, the mean value of the kinetic energy of an orbiting object is half that of the mean value of its potential energy, that is,

$$
\langle T \rangle = \frac{1}{2}\langle V \rangle,
$$

from which it follows that,

$$
\oint \frac{GM}{h}dt = \oint \frac{GM}{a}dt,
$$

giving again as required,

$$
\Delta \varepsilon_{twin} = \Delta \varepsilon_{Newton} = -\frac{GM}{2a}.
$$

4   The Schwarzschild metric

The Schwarzschild metric is a spherically symmetric and time-independent solution to Einstein's field equation in a vacuum. It may be written (see, for example, Misner et al [13], at p. 607),

$$ds^2 = -d\tau^2 = -\left(1 - \frac{2GM}{r}\right)dt^2 + \left(1 - \frac{2GM}{r}\right)^{-1}dr^2$$
$$+ r^2 d\theta^2 + r^2 \sin^2 \theta d\phi^2 \qquad (15)$$

where,

ds is the invariant interval,

$d\tau$ is proper time – the time actually experienced within the metric,

dt is coordinate time – the time experienced by a notional observer at infinity,

r, θ and ϕ are spherical coordinates,

G is again the universal gravitational constant, and

M is a central mass.

The successes of the Schwarzschild metric include the anomalous precession of the perihelion of Mercury, the gravitational deflection and lensing of light, the redshift of light, and the Shapiro delay, see generally Will [14].

As discussed in Sect. 1, a distinction is commonly drawn in discussions of global positioning systems between time dilations due to special relativity (the twin effect) and those attributed to general relativity [4]. Drawing the same distinction here, the contribution from special relativity can be isolated by ignoring temporarily the central mass M, whereupon the metric (15) becomes,

$$d\tau^2 = dt^2 - dr^2 - r^2 d\theta^2 - r^2 \sin^2 \theta \, d\phi^2, \qquad (16)$$

which is the metric of Minkowski spacetime expressed in spherical coordinates.

For an object making a return trip, this metric must induce the twin effect, as can be verified by expressing velocity in those same spherical coordinates, that is,

$$v_r = \frac{dr}{dt},$$
$$v_\theta = r\frac{d\theta}{dt},$$
$$v_\phi = r \sin \theta \frac{d\phi}{dt}.$$

---

[4]When considered from beyond the effects of gravity, these dilations are cumulative. But when considered from the surface of the Earth, the dilation due to gravity is less at the satellites than on the ground. The net effect is that the clocks of the satellites run faster than those on the ground, see Ashby [4].

and using these expressions to eliminate dr, $d\theta$ and $d\phi$ from Eqn. (16), which becomes,

$$d\tau^2 = dt^2 \ (1 - v_r^2 - v_\theta^2 - v_\phi^2) = dt^2 \ (1 - v^2),$$

giving for a complete orbit, as expected, the twin effect,

$$\tau = \oint (1 - v^2)^{\frac{1}{2}} \, dt. \tag{17}$$

Eqn. (17) will provide the dilation of time due to the twin effect for any return trip at all, including a meandering (and thus powered) excursion. But for a system of mutually orbiting objects, specifically here a test particle orbiting a much larger central mass, the trajectories of interest are those in which there is no variation in either the energy or the angular momentum of the orbiting masses.

In such a system, orbital motion once induced must endure, and my objective in what follows will be to show that in accommodating the further constraints imposed by conservation and the principle of relativity, the flat space metric of Eqn. (16) becomes the Schwarzschild metric of Eqn. (15). I will consider those constraints in the next section, postponing to Sect. 8, the consideration of why the predictions of the metric should approximate those of an inverse square law[5].

5   The metric components g_{tt} and g_{rr}

As can be seen by comparing Eqns. (15) and (16), it is only in the tensor components,

$$g_{tt} = \left(1 - \frac{2GM}{r}\right), \ and$$

$$g_{rr} = \left(1 - \frac{2GM}{r}\right)^{-1},$$

(which describe, respectively, a dilation of time and an expansion of radial distance), that the Schwarzschild metric differs from that of Minkowski.

---

[5]Nor do I consider here why G has the value it has, other than to suggest that if gravity is a consequence of the twin effect, G may not be fundamental, but a measure of the degree to which matter is gravitationally bound in the present epoch. In that case, G would be akin to an intensive thermodynamic parameter, its apparently unchanging current value having been determined by the circumstances of the early universe.

That some such dilation of time should be expected may be inferred by noticing that since an object has a diminished energy in a gravitational potential, it must also have the correspondingly reduced frequency contemplated by the Planck-Einstein relation (1). But before giving further consideration to the value of g_{tt}, it will be helpful to discuss the constraints imposed by the principle of relativity on the relationship between g_{tt} and g_{rr}.

Notice firstly that if a particle experiences a reduction of frequency as a result of its depth within a gravitational potential, so also in the same degree must every other particle at the same depth, including every photon that is emitted at that depth when a system transitions from one state to another. If that were not so, inter-particle interactions would not be consistently the same at that depth as stipulated by the principle of relativity.

As frequencies are reduced and time becomes dilated in the same degree for all physical processes, including biophysical and mental processes, time itself runs more slowly within the gravitational field than it does outside it (or at least will seem to do so).

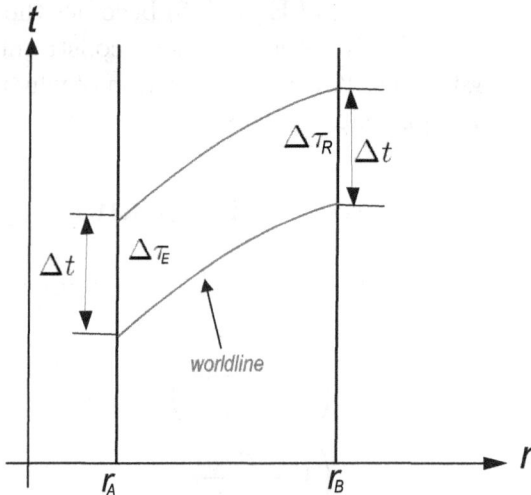

Figure 2: The worldlines of successive crests of an electromagnetic wave emitted at r_A and received at r_B. The coordinate time Δt between departures from r_A must be the same as that between corresponding arrivals at r_B. But the proper times will differ.

What then of wavelengths? Consider Fig. 2, which is a space-time diagram of a kind commonly found in discussions of gravitational redshifts, see for instance Zee [15], at p. 304, and Moore [16], at p. 109. The diagram depicts the worldlines of successive crests of an electromagnetic wave emitted at r_A and received at r_B. Because these

166

successive paths from r_A to r_B are congruent, the coordinate time Δt between departures from r_A must be the same as that between corresponding arrivals at r_B. From the standpoint of a notional observer beyond the influence of the gravitating mass (the notional observer at infinity) the wave will thus have the frequency at r_B that it had at r_A[6].

However the corresponding proper times will differ. The time actually experienced by a particle or observer at r_A will be be less than at r_B. An incoming photon from a transition at r_A will have a frequency less than that of a photon emitted from the corresponding transition at r_B.

Now consider what this implies for wavelengths. If, as assumed by special relativity, the photon from r_A is to be observed at r_B to have the velocity c, it follows from the relation,

$$\frac{\omega}{k} = c,$$

that it must have a correspondingly longer wavelength,

$$\lambda = \frac{2\pi}{\kappa},$$

than the photon emitted from the same process at r_B, which means in effect that lengths must increase as times become dilated, that is,

$$g_{rr} = g_{tt}^{-1}. \tag{18}$$

It is commonly, but rather loosely, said in discussions of the gravitational redshift, (see, for instance Wald [17], at p. 137) that a photon suffers a loss of energy and reduction of frequency as it rises from a gravitating mass (the redshift) and gains energy and increases in frequency as it falls toward the gravitating mass (the blueshift). But as can be seen from the discussion above, the incoming photon is observed to be redshifted or blueshifted, as the case may be, because it was created in a reference frame in which frequencies, energies and wavelengths differ from those in the reference frame in which it is subsequently measured. If the photon were in fact to gain or lose energy, this would be contrary to the law of conservation of energy.

g_{tt} can now be deduced by considering the energy lost by a test particle that, after falling vertically from infinity, is brought to a stop

---

[6]A t-meter, may aid the understanding here, see Moore [16], at p. 108.

at a distance r from the centre of the mass M. This loss must be evaluated in the r and t coordinates of the Schwarzschild metric, and I will assume here (and discuss further in Sect. 8) that it is also in these coordinates that the inverse square law applies, this being why in this metric,

$$V(r) = -\frac{GM}{r},$$

(see Moore [16], at p. 117). On that assumption, a particle that has fallen from infinity has at r the velocity,

$$v(r, t) = \frac{dr}{dt} = (\frac{2GM}{r})^{\frac{1}{2}}. \tag{19}$$

As the particle falls, and for as long as its fall remains uninterrupted, its energy remains constant, retaining the value, $E_0 = m_0$, that it had at infinity. It thus follows from the relativistic equation of motion,

$$E^2 - p^2 = m^2, \tag{20}$$

that as the momentum p of the particle increases, its relativistic (or effective) mass must decrease[7]. Let us suppose then that m decreases with r in accordance with an as yet unknown function $f(r)$, so that we can write,

$$m \rightarrow m_0 f,$$
$$p \rightarrow \frac{m_0 f v}{(1 - v^2)^{\frac{1}{2}}},$$

and the relativistic equation of motion becomes,

$$(m_0)^2 - \frac{(m_0 f v)^2}{1 - v^2} = (m_0 f)^2,$$

whereupon, on solving for f, we have,

$$f = (1 - v^2)^{\frac{1}{2}}.$$

When brought to a stop at r, the particle loses its momentum and is left with the energy,

$$E = (1 - v^2)^{\frac{1}{2}} E_0,$$

---

[7]The notion of a varying relativistic mass seems indispensable here, see generally Petkov [18], Chap. 9.

so that from Eqn. (19), we have, as required,

$$g_{tt} = (\frac{d\tau}{dt})^2 = (1 - \frac{2GM}{r}),$$

while from Eqn. (18),

$$g_{rr} = (1 - \frac{2GM}{r})^{-1}.$$

One question remains: why should a moving particle experience, both the dilation of time defined for a moving particle by the twin effect, as well as the dilation of time experienced by a stationary particle at the same position? The answer, shortly stated, is that the principle of relativity demands it. The laws of physics must hold in the same manner at r as they do for a particle that is beyond the reach of gravity, which could not be the case if a particle moving at r had a rest mass that differed from that of the stationary particle at the same position.

6   The de Broglie wave

The Schwarzschild metric has thus been explained from the twin effect, which as shown in Sect. 2 can itself be explained from the evolution of phase defined by the de Broglie wave. As discussed in Sect. 1, this same evolution of phase provided the basis for quantum mechanics, which raises the possibility that gravity and quantum mechanics have a common origin in wave structure.

But the de Broglie wave is superluminal and not as yet a suitable candidate for reconciling these two theories. What must first be provided is a physically reasonable provenance for the de Broglie wave itself.

I will argue that the de Broglie wave is better understood, not as an independent wave, but as the modulation (or dephasing or beating, see, for instance, Feynman et al [19], Vol. I, Chap. 48), of an underlying wave structure that is itself evolving through space at the subluminal velocity v of the particle. I will base this argument on two empirically well-established laws of physics, the Lorentz transformation and the Planck-Einstein relation (Eqn. (1) above).

Before presenting this argument, I should first explain that this way of understanding the de Broglie wave is not at all novel. The de Broglie wave can be seen to emerge in the manner to be now described in two of the three demonstrations of the de Broglie wave in de Broglie's own thesis of 1923 [20], one being a treatment in Minkowski spacetime [20], Chap. I, Sect. III, and the other, an intuitively more accessible mechanical model comprising an array of oscillating springs [20], Chap. I, Sect. II. This interpretation of the de Broglie wave was subsequently noticed by Mellen [21], and discussed at length by Wolff [22]. It has now acquired a "literature", as listed in Shanahan [23], and see also Shanahan [24] to [26].

On the evidence of the Lorentz transformation, a massive particle comprises in its entirety underlying influences evolving at the velocity c of light. If that were not so – if there were some other velocity having the same fundamental significance as c – such a velocity would have its own Lorentz factor γ (see Eqn. (3)), and its own Lorentz transformation based on that factor γ, and the laws of physics would not then be the same from one inertial frame to the next. While a particle comprising underlying influences of more than one fundamental velocity might be stable in one inertial frame of reference, it could not retain the stability of its characteristic structure in any other inertial frame.

There are, of course, velocities in Nature that differ from c, those for instance of massive particles, sound waves, and refracted light. But, as Einstein explained in 1905 [27], such velocities transform in accordance with the relativistic formula for the composition of velocities. To explain the all-encompassing ambit of the Lorentz transformation, these other velocities must be regarded, not as fundamental, but as existentially dependent on c, that is to say, as the net effect of underlying influences that *do* evolve at velocity c.

On the evidence of the Planck-Einstein relation (Eqn. (1)), it should also be assumed that these underlying influences of velocity c are wave-like in nature, and that in its rest frame, a massive particle comprises wave-like influences having the characteristic frequency ω_0 of the species of particle in question. In consequence, the particle would then have a characteristic wave number κ_o, satisfying the relation,

$$\frac{\omega_0}{\kappa_0} = c,$$

and thus a corresponding wavelength,

$$\lambda_0 = \frac{2\pi}{\kappa_0},$$

thereby according physical meaning to the Compton wavelength,

$$\lambda_c = \frac{2\pi}{k_0} = \frac{h}{mc}.$$

170

There is a wealth of corroborating evidence for this wave-based understanding of the nature of solid matter, but the item of evidence that seems particularly compelling is (as I will now show) the origin it provides for the de Broglie wave.

If, as argued above, a massive particle comprises underlying wave-like influences of velocity c, it must comprise in its rest frame some form of standing or stationary wave. It is not necessary to consider the details of such a structure for it is easily shown that every standing or stationary waveform gives rise to a de Broglie wave when considered from another inertial frame of reference.

Consider the standing wave,

$$R(x, y, z)\, e^{i\omega t}, \tag{21}$$

which is evolving in time at some frequency ω, but for which no assumption has been made as to its manner of spatial variation. Under a relativistic boost in the x-direction, this waveform becomes the moving wave,

$$R(\gamma\,(x - vt),\, y,\, z)\, e^{i\omega\gamma(t - vx)}. \tag{22}$$

in which the spatial factor $R(x, y, z)$ of standing wave (21) has become the carrier wave,

$$R(\gamma\,(x - vt),\, y,\, z), \tag{23}$$

which is evidently moving through space at the velocity v and, as indicated by the presence of the Lorentz factor γ, is exhibiting the contraction of length predicted by special relativity.

The second factor,

$$e^{i\omega\gamma(t - vx)}, \tag{24}$$

in wave (22) is a transverse plane wave, which (in these units where $c = 1$ and $v < c$) is evolving through the carrier wave (23) at the superluminal velocity v^{-1}. If the frequency ω is now identified as the natural frequency ω_0 of a massive particle, wave factor (24) can be rewritten in terms of the Einstein frequency,

$$\omega_E = \frac{E}{\hbar}\gamma\omega_0, \tag{25}$$

and de Broglie wave number,

$$\kappa_{dB} = \frac{p}{\hbar} = \gamma\omega_0\, v, $$

as,

$$e^{i(\omega_E\, t - \kappa_{dB}\, x)}, \tag{26}$$

171

and is now recognizable as the de Broglie wave, no longer an independent wave, but a modulation or dephasing of the underlying carrier wave. From Eqns. (22) and (26), the full composite particle wave structure is then,

$$R(\gamma\,(x-vt)\,,\,y,\,z)\,e^{i(\omega_E t - \kappa_{dB} x)}. \tag{27}$$

In this interpretation, this otherwise anomalous superluminal phenomenon achieves consistency with special relativity. Its superluminal velocity is no longer an embarrassment as the velocity of a modulation is not that of energy or information transport. And unlike the de Broglie wave considered alone, the full modulated wave structure (27) is a manifestly covariant relativistic object, capable in principle of taking its place in the tensor equations of relativistic physics. The Fitzgerald-Lorentz contraction appears in the carrier wave (23), while the dilation of time and failure of simultaneity are described by the modulation, that is to say, by the de Broglie wave (26).

The effect of the modulation is that the various parts of the moving wave are no longer cresting in unison as they had been in the standing wave, but in sequence, those ahead lagging in phase (and thus time) those behind. The de Broglie wave describes a progressive loss of phase in the direction of travel corresponding exactly in effect to the failure of simultaneity in that direction predicted by the Lorentz transformation.

Once the existence of the underlying wave structure is recognized, several mysteries become resolved of which I will mention below only those of relevance to the reconciliation of gravity and quantum mechanics.

7   Quantum mechanics from particle wave structure

If a massive particle were some form of tiny solid object, it would be exceedingly curious that the energy E and momentum p of this object should be associated with the wave characteristics, ω and κ respectively, of a superluminal wave with which it would seem to have no physical nexus. Yet it was essentially on the basis of that association that wave mechanics was originated by Schrödinger and has since developed.

In constructing a wave equation that would have solutions consistent with the Planck-Einstein and de Broglie relations, Eqns. (1) and

(9) respectively, Schrödinger made the substitutions,

$$p \to i\hbar \frac{\partial}{\partial x}, \text{ and}$$

$$E \to i\hbar \frac{\partial}{\partial t},$$

in the non-relativistic equation of motion,

$$E^2 = \frac{p^2}{2} + V,$$

to obtain the non-relativistic Schrödinger equation,

$$i\hbar \frac{\partial \psi}{\partial t} = -\frac{\hbar^2}{2m} \nabla^2 \psi + V\psi,$$

as he had likewise done in the corresponding relativistic equation of motion to obtain the corresponding relativistic wave equation (now called the Klein-Gordon equation).

These equations have owed their utility to the correspondence between, on the one hand, the frequency and wave number of the de Broglie wave, and on the other, the energy and momentum, respectively, of the associated particle. But it is the existence of the underlying carrier wave that makes sense of that correspondence, and it is the full wave (27), rather than the de Broglie wave considered alone, that provides an understanding of the nature of mass, energy, momentum and inertia.

In order to show that this is so, it will be helpful to have before us a model that, unlike the model described by Eqn. (27), displays the underlying velocity c. I will take as a suitable model,

$$\psi(\mathbf{r}, t) = \frac{1}{2} |\mathbf{r}|^{-1} [e^{i(\omega_o t - \check{\kappa}_o \cdot \mathbf{r})} - e^{i(\omega_o t + \check{\kappa}_o \cdot \mathbf{r})}], \tag{28}$$

which has the idealized form of a spherical standing wave, centred at $\mathbf{r} = 0$ and constructed from incoming and outgoing waves of velocity c, where,

$$\frac{\omega_o}{\kappa_o} = c,$$

κ_o being here the Compton wave number.

Model wave (28) has a singularity at the origin and is itself unphysical, but will illustrate how the dynamic properties of a massive particle might originate from a thoroughly wave-theoretic treatment of matter.

On a boost in the x-direction, wave (28) becomes (on taking real parts),

$$\Psi(x, y, z, t) = \sin \kappa_o \sqrt{\gamma^2(x - vt)^2 + y^2 + z^2} \, \cos(\omega_E t - \kappa_{dB} x), \quad (29)$$

(where an amplitude factor has been omitted).

As in the case of travelling wave (27), wave (29) comprises, as one factor, a carrier wave of velocity v,

$$\sin \kappa_o \sqrt{\gamma^2(x - vt)^2 + y^2 + z^2}, \quad (30)$$

and as modulating factor, the de Broglie wave,

$$\cos(\omega_E t - \kappa_{dB} x),$$

which is of planar form and is evolving through the carrier wave at the superluminal velocity v^{-1}.

To show how this modulated wave structure is related to dynamic changes in the particle, it will suffice to consider rays passing through the centre of the waveform and moving forwardly and rearwardly along the direction of travel[8]. In the rest frame of the particle, the superposition of these rays produces the one-dimensional standing wave,

$$\Psi(x, t) = \frac{1}{2}[e^{i(\omega_o t - \kappa_o x)} - e^{i(\omega_o t + \kappa_o x)}], \quad (31)$$

but when observed from a frame in which the particle is moving at velocity v, these forwardly and rearwardly moving rays, to be now labelled 1 and 2 respectively, transform as,

$$e^{i(\omega_o t - \kappa_o x)} \quad \rightarrow \quad e^{i(\omega_1 t - \kappa_1 x)},$$
$$e^{i(\omega_o t + \kappa_o x)} \quad \rightarrow \quad e^{i(\omega_2 t + \kappa_2 x)},$$

where, in accordance with the Doppler effect,

$$\omega_1 = \gamma \omega_0 (1 + v), \quad \omega_2 = \gamma \omega_0 (1 - v), \quad (32)$$
$$\kappa_1 = \gamma \kappa_0 (1 + v), \quad \kappa_2 = \gamma \kappa_0 (1 - v), \quad (33)$$

so that standing wave (31) becomes,

$$\Psi(x, t) = \frac{1}{2}[e^{i(\omega_1 t - \kappa_1 x)} - e^{i(\omega_2 t + \kappa_2 x)}],$$

---

[8]For a consideration of rays in other directions, see Shanahan [24].

which can also be written,

$$\Psi(x,t) = \sin\left(\frac{\omega_1 - \omega_2}{2}t - \frac{\kappa_1 + \kappa_2}{2}x\right)\cos\left(\frac{\omega_1 + \omega_2}{2}t - \frac{\kappa_1 - \kappa_2}{2}x\right).$$
(34)

In wave (34), the second factor,

$$\cos\left(\frac{\omega_1 + \omega_2}{2}t - \frac{\kappa_1 - \kappa_2}{2}x\right),$$

is the de Broglie wave, from which the Einstein frequency and de Broglie wave number are therefore, respectively,

$$\omega_E = \frac{\omega_1 + \omega_2}{2},$$

and,

$$\kappa_{dB} = \frac{\kappa_1 - \kappa_2}{2}.$$

In natural units in which $\hbar = c = 1$, the Planck–Einstein relation (1) is thus,

$$E = \frac{\omega_1 + \omega_2}{2},$$
(35)

while the de Broglie relation (9) is simply,

$$p = \frac{\omega_1 - \omega_2}{2}.$$
(36)

The energy and momentum of the particle have thus been expressed in a simple and intuitive way as, respectively, the sum of and the difference between, the energies of forwardly and rearwardly moving waves.

In the same natural units, it follows from Eqns. (32) and (33) that,

$$m = \omega_0 = \sqrt{\omega_1 \omega_2},$$
(37)

while the relativistic equation of motion (Eqn. (20), that is,

$$E^2 - p^2 = m^2,$$

can be seen to be the equality,

$$\left(\frac{\omega_1 + \omega_2}{2}\right)^2 - \left(\frac{\omega_1 - \omega_2}{2}\right)^2 = \omega_0^2.$$
(38)

If inertia is now interpreted, not simply as the resistance of a massive particle to changes in its state of motion, but at a more fundamental level, as the resistance of a wave to changes in its oscillatory state, we have in Eqns. (35) to (38), a consistent scheme for the treatment in terms of wave characteristics of the energy, momentum, inertia and mass of a massive particle.

In summary, it has been argued in this and the preceding section: that on the evidence of the Lorentz transformation and Planck-Einstein relation, the elementary articles comprise underlying wave-like influences of velocity c; that a massive particle must therefore comprise in its rest frame some form of standing wave; that when observed from another inertial frame this standing wave becomes a travelling wave from which the de Broglie wave emerges as a modulation; and that when considered in this way, the dynamic properties of a massive particle become the properties of a waveform.

8 Gravity from particle wave structure

With the de Broglie wave explained, not as an improbable wave of superluminal velocity, but in a manner well-known from wave-theory (see again Feynman et al [19], Vol. I, Chap. 48), it is now possible to present a wave-theoretic explanation of the twin effect in which gravity emerges from the manner in which this wave structure must adapt to a change of inertial frame.

In Fig. 3, a test particle P is depicted at three locations along its elliptical path about a central body O. To a co-moving observer, P would have the form of a standing wave attenuated in intensity in accordance with an inverse square law. But to a notional observer at O, the wave structure of P is continually (and continuously) adapting to its orbital path in the manner described for model particle (28) by Eqn. (29). As observed from O, the underlying carrier wave is contracted in the direction of motion, whilst its modulation (the de Broglie wave), which has been represented in the drawing by parallel transverse lines, is evolving through the ellipsoidally-contracted wavefronts of the carrier wave at superluminal velocity.

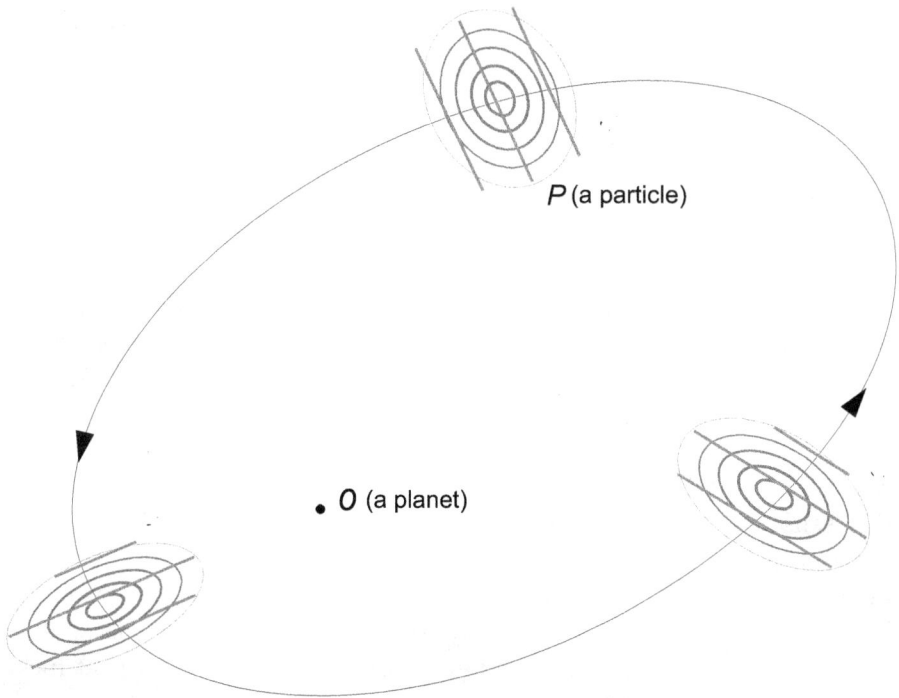

Figure 3: The wave structure of a massive particle P depicted in a symbolic manner at three positions along its orbit about a central mass O.

Considered, not as a wave in its own right, but as the modulation of an underlying wave structure, it is no longer puzzling that the superluminal de Broglie wave does not fly off at a tangent from its orbital path, a difficulty that confounded attempts, notably by Schrödinger (see Dorling [28]), to fit this mysterious "wave" to the orbits of atomic electrons. Nor should it now seem mysterious that the de Broglie wave appears to be "piloting" the subluminal particle along its path, and yet at the same time, is continually overtaking but never leaving the moving particle. In effect, carrier wave and de Broglie modulation are wave factors in the one integral whole.

But the effect of relevance to gravity here is that, as a consequence of the modulation, the orbiting particle experiences the loss of phase in the direction of travel that results in the reduction of energy per orbit that was shown above to reproduce the binding energies of Newtonian gravity and the Schwarzschild metric.

While Newton was able to show that the closed elliptical orbits described by Kepler imply a force acting directly between one mass and another, he was unable to explain the origin of such a force. Nor should it be thought that the curvature of spacetime supposed by general relativity is an adequate explanation of gravity. There is, of course, a pleasing symmetry in Wheeler's aphorism that,

> Space acts on matter, telling it how to move. In turn, matter reacts back on space, telling it how to curve (Misner et al [13], p. 5).

But the geometric approach provides no apparent basis for this reciprocity of cause and effect. While the geodesic equation does explain how matter would move in a curved spacetime, general relativity is silent as to how matter causes this curvature.

On the other hand, in the interpretation of gravity proposed in this paper, the movement of matter provides its own explanation of why an object must persist in a bound state. Gravity becomes, in this sense, a form of inertia. Einstein stressed the close relationship between gravity and inertia, and on occasion referred to gravity as being a form of inertia, see Lehmkuhl [30]. In the scheme proposed here, it is likewise the inertia of a system of mutually orbiting objects – the tendency of the system to persist in its present state – that locks in the mass defect of the system and ensures that, failing the loss or input of further energy, the system retains that state.

But why should this binding effect follow (as it does exactly in Newtonian gravity) an inverse square law. Such a law is commonly explained from the geometric dilution of an effect that radiates outwardly from a point (see, for instance Wikipedia [29]). A similar dilution must occur in the wave-based explanation of matter described above, where the amplitude of a particle wave structure must decrease inversely with radial distance. Were it not to do so there would be a discontinuity in the movement of energy inwardly and outwardly through the wave structure.

Assuming that this same requirement of continuity constrains the composite wave structure of an aggregation of particles, this would explain the inverse square law governing Newtonian gravity. It would also be consistent with the apparent departure from the predictions of that law in the Schwarzschild metric, where the effective rest mass of the particle decreases with depth within the potential.

9   Concluding remarks

It would be only natural to suggest that I have the cart before the horse here – that in the ordering of explanatory priority, it is not the twin effect that explains gravity, but gravity that is the cause of orbital motion and is thus the explanation of the twin effect.

But it is not gravity, but the failure of simultaneity described by the Lorentz transformation that is the source of the twin effect, and that transformation applies to all matter, whether moving rectilinearly, following an orbital path, or moving in any other way. I stress again that even if there were no gravity in the usually recognized sense, the Lorentz transformation would induce the twin effect and an orbiting object would experience a dilation of time, and have nonetheless, a binding energy.

There is also here the promise of explanatory unification. On the evidence of the Planck-Einstein relation and Lorentz transformation, and with the corroboration of the de Broglie wave, I have argued for a unified wave-theoretic understanding of matter and radiation, implying in turn a common origin in wave structure for gravity and quantum mechanics.

There is thus, I suggest, ample reason to decide the issue of explanatory priority in favour of the twin effect, or to take this to a more fundamental level, in favour of a thoroughly wave-based explanation of matter and energy.

References

[1] A. Eddington, Space, Time and Gravitation: An outline of the general relativity theory, Cambridge University Press, Cambridge U. K. (1920)

[2] J. C. Hafele, R. E. Keating, Around-the-World Atomic Clocks: Observed Relativistic Time Gains, Science **177**, 168 (1972)

[3] M. P. Balandin, V. M. Grebenyuk, V. G. Zinov, A. D. Konin, A. N. Pononmarev, Measurement of the lifetime of the positive muon, Sov. Phys. JETP. 40, 811 (1974)

[4] N. Ashby, Relativity and the Global Positioning System, Physics Today, May, 2002

[5] F. Bloch, Heisenberg and the early days of quantum mechanics, Physics Today, Dec. 1976, adapted from a talk given on 26 April 1976 at the Washington D. C. meeting of The American Physical Society

[6] P. A. M. Dirac, The physical interpretation of the quantum dynamics, Proc. Roy. Soc. London A **113**, 621 (1927)

[7] R. Bate, D. Mueller, J. White, Fundamentals of Astrodynamics, Dover, New York (1971)

[8] H. Minkowski, Raum und Zeit, Phys. Zeits. **10**, 104 (1909), English trans., Space and Time, by F. Lewertoff and V. Petkov, in V. Petkov (ed.), Space and Time, Minkowski's Papers on Relativity, Minkowski Institute Press, Montreal, Quebec (2012)

[9] I. Newton, The Principia, Mathematical Principles of Natural Philosophy, as translated by I. B. Cohen, A. Whitman, assisted by J. Budenz, preceded by a guide to the Principia by I. B. Cohen, University of California Press, Berkeley (1999)

[10] I. B. Cohen (ed.) Isaac Newton's Papers & Letters on Natural Philosophy and Related Documents, Harvard University Press (1958)

[11] T. Logsdon, Orbital Mechanics: Theory and Applications, Wiley, Hoboken, New Jersey (1997)

[12] H. Goldstein, Classical Mechanics, 2nd Ed., Addison-Wesley, Reading, Mass. (1980)

[13] C. W. Misner, K. S. Thorne, J. A. Wheeler, Gravitation, W. H. Freeman, New York (1973)

[14] C. M. Will, Was Einstein Right?, Oxford University Press, Oxford (1986)

[15] A. Zee, Einstein Gravity in a Nutshell, Princeton University Press, Princeton (2013)

[16] T. A. Moore, A General Relativity Workbook, University Science Books, Mill Valley, California (2013)

[17] R. M. Wald, General Relativity, University of Chicago Press, Chicago (1984)

[18] V. Petkov, Relativity and the Nature of Spacetime, 2nd. Ed., Springer, Heidelberg (2009)

[19] R. P. Feynman, R. B. Leighton, M. Sands, The Feynman Lectures on Physics. Addison-Wesley, Reading (1963)

[20] L. de Broglie, Doctoral thesis. Recherches sur la théorie des quanta. Ann. de Phys. (10) **3**, 22 (1925).

[21] W. R. Mellen, Moving Standing Wave and de Broglie Type Wavelength, Am. J. Phys. **41**, 290 (1973)

[22] M. Wolff, The Wave Structure of Matter and the origin of the Natural Laws, in R. L. Amoroso, G. Hunter, M. Kafatos, J.-P. Vigier (eds.), Gravitation and Cosmology: From the Hubble Radius to the Planck Scale, Kluwer, Dordrecht (2002)

[23] D. Shanahan, Reverse Engineering the de Broglie Wave IJQF **9**, 44-63 (2023)

[24] D. Shanahan, A Case for Lorentzian Relativity, Found. Phys. **44**, 349 (2014)

[25] D. Shanahan, The de Broglie Wave as Evidence of a Deeper wave Structure arXiv:1503.02534v2 [physics.hist-ph]

[26] D. Shanahan, What might the matter wave be telling us of the nature of matter? IJQF **5**,165 (2019)

[27] A. Einstein, Zur elektrodynamik bewegter Korper Ann. Phys. **17**, 891 (1905), English trans., On the electrodynamics of moving bodies, in H. A. Lorentz, A. Einstein, H. Minkowski, H. Weyl, The Principle of Relativity, Methuen, London (1923)

[28] J. Dorling, Schrödinger's original interpretation of the Schrödinger equation: a rescue attempt, in C. Kilmister (ed.), Schrödinger, centenary celebration of a polymath, Cambridge University Press, Cambridge (1987)

[29] Inverse-square law, In Wikipedia, Retrieved, Jan. 3 2023. http:wikipedia.org/wiki/Inverse-square law

[30] D. Lehmkuhl, Why Einstein did not believe that general relativity geometrizes gravity, Studies in History and Philosophy of Modern Physics, **46**, 316-326 (2014)

9 ON THE SPECIAL PRINCIPLE OF RELATIVITY

HIROAKI FUJIMORI

Abstract Looking back on the history of the special theory of relativity, Lorentz and Poincaré were on their way to giving a mathematical proof to the result of Michelson Morley experiment. Meanwhile Einstein published the theory of relativity based on the principle of relativity and the constancy of the speed of light [1]. The problem left by Poincaré in his book "Électricité et optique" [2] was that a well-developed theory should be able to prove this principle [of relativity] very strictly and all at once. By the way what do you do when you prove that Euclidean geometry holds completely the same on both the front and back symmetric plane? Here is the answer to this problem. By establishing a mathematical derivation of the special principle of relativity, my paper gives a positive answer to this problem.

Keywords: special principle of relativity, Curie's principle, Lorentz transformation

1 Preliminaries

1.1 Terms

- *Spacetime* is a four-dimensional unified entity of space and time without considering all of the matter from the universe. The 4-D *spacetime* is also known as Minkowski space.
- *Symmetry plane* is a plane with indistinguishable back and front surfaces.
- Symmetry plane in spacetime is one of two types, namely, *space × space* type or *space × time* type. A *space × space* type plane is completely isotropic, since the space line is isotropic. A *space × time* type plane is semi-isotropic, since the time line is of unidirectional

A. S. Stefanov, G. Dupuis-Mc Donald (Eds), *Spacetime Conference - 2022*. *Selected peer-reviewed papers presented at the Sixth International Conference on the Nature and Ontology of Spacetime, 12 - 15 September 2022, Albena, Bulgaria* (Minkowski Institute Press, Montreal 2023). ISBN 978-1-989970-96-6 (softcover), ISBN 978-1-989970-97-3 (ebook).

and the space line is isotropic. This type of a plane is also known as Minkowski plane.

1.2 Special principle of relativity

Extended Curie's principle

Curie's principle is that in linear physical phenomena, the symmetries of the causes are to be found in the effects. We have extended this principle that in linear space, the symmetry of space is to be derived from the symmetries of subspaces, so is the spacetime transformation. This means that both basic symmetries of isotropy of space and unidirectional of time should be used in order to derive the spacetime transformation.

From this viewpoint, when we review the methodology of the special relativity, the basic symmetry of unidirectional of time which is two different time lines have the same directions of time is not used explicitly in any process on deriving Lorentz transformation, while another basic symmetry of isotropy of space is used explicitly. The condition of unidirectional of time is hidden behind the principles. When we think that the principle of the constancy of the speed of light depends on both the principle of relativity and the Maxwell's law of the speed of light, the remaining independent principle of relativity must be a sufficient but not necessary condition. This might be what Poincaré minded.

Definition of the special principle of relativity

Let $\boldsymbol{A} = \begin{pmatrix} a & b \\ c & d \end{pmatrix}$ be a linear transformation, $f(p)$ and $g(q)$ be functions, $p = \begin{pmatrix} x \\ y \end{pmatrix}$ and $q = \begin{pmatrix} u \\ v \end{pmatrix}$ be points on different right-hand coordinate planes, and $q = \boldsymbol{A}p$, $f(p) = f\left(\boldsymbol{A}^{-1}q\right) := g(q)$ by definition hold. If $g(q) = f(q) = f(p)$, then we are not able to distinguish which world of the coordinate plane we are on. This shows how a law $f(p) = f(q)$ is born which means the same function f on the different coordinate plane of p and q, and this mechanism is called the principle of relativity on physics, which is the same as the thought of Erlangen program on geometry. This principle (law) must be subject to the extended Curie's principle. The mathematical definition of the principle of relativity is $\underline{f(q) = f(\boldsymbol{A}p) = f(p)}$ and $\underline{\det \boldsymbol{A} > 0}$.

1.3 Theorems[1]

■ *Invariant line of a 2 × 2 matrix*

When the matrix $\boldsymbol{B} = \begin{pmatrix} a & b \\ c & d \end{pmatrix}$ has an eigen value $\lambda = 1$, it has an invariant line f(p).

$$\mathrm{f}(\boldsymbol{B}p) \equiv \mathrm{f}(p) = cx - (a-1)y. \tag{1}$$

■ *Identity of quadratic invariant function of a 2 × 2 matrix*

Let $\boldsymbol{A} = \begin{pmatrix} a & b \\ c & d \end{pmatrix}$ be a matrix, $p = \begin{pmatrix} x \\ y \end{pmatrix}$ be a point, and $\phi(p) = -cx^2 + by^2 + (a-d)xy$ be a quadratic function. There exists an identity $\phi(\boldsymbol{A}p) \equiv \det \boldsymbol{A} \cdot \phi(p)$.

If $\det \boldsymbol{A} = 1$, then $\phi(\boldsymbol{A}p) = \phi(p)$. $\tag{2}$

In this case, $\phi(p)$ is called a quadratic invariant function of transformation $\boldsymbol{A}$.

Note that this relation satisfies the principle of relativity.

■ *Polar form of a 2 × 2 special linear matrix*

A special linear matrix $\boldsymbol{S}$ is decomposed using commutative coefficients k, h such as

$$\boldsymbol{S} = \begin{pmatrix} a & b \\ c & d \end{pmatrix} = \begin{pmatrix} m + hb & b \\ kb & m - hb \end{pmatrix}, \tag{3}$$

where $\det \boldsymbol{S} = m^2 - \Delta b^2 = 1, \Delta = h^2 + k, m = (a+d)/2, k = c/b, 2h = (a-d)/b, b \neq 0$.

The matrix $\boldsymbol{S}$ has a normalized invariant function as given by

$$\phi(\boldsymbol{S}p) = \phi(p) = -kx^2 + y^2 + 2hxy = r^2, \quad \det \boldsymbol{S} = 1, p = \begin{pmatrix} x \\ y \end{pmatrix}. \tag{4}$$

The special linear transformation $\boldsymbol{S}$ satisfies the definition of the principle of relativity. The matrix $\boldsymbol{S}$ and the invariant function $\phi(p)$ is classified into three types based on the sign of the discriminant $\Delta = h^2 + k$. If $\Delta < 0$, then they are of an elliptic type. If $\Delta > 0$, then they are of a hyperbolic type. If $\Delta = 0$, then they are of a linear type. Thus we represent a 2 × 2 special linear matrix $\boldsymbol{S}$ ($\det \boldsymbol{S} = m^2 - \Delta b^2 = 1$) by *declination* θ and commutative coefficients k, h, using the relations of $\cos^2 \theta + \sin^2 \theta = 1$ or $\cosh^2 \theta - \sinh^2 \theta = 1$.

---

[1]These original theorems have proofs in the book [6] or paper [7].

For $\Delta < 0$ (elliptic type),

$$S = S(\theta, k, h) = \begin{pmatrix} \cos\theta + \frac{h}{\sqrt{-\Delta}}\sin\theta & \frac{1}{\sqrt{-\Delta}}\sin\theta \\ \frac{k}{\sqrt{-\Delta}}\sin\theta & \cos\theta - \frac{h}{\sqrt{-\Delta}}\sin\theta \end{pmatrix}. \tag{5}$$

For $\Delta > 0$ (hyperbolic type),

$$S = S(\theta, k, h) = \begin{pmatrix} \cosh\theta + \frac{h}{\sqrt{\Delta}}\sinh\theta & \frac{1}{\sqrt{\Delta}}\sinh\theta \\ \frac{k}{\sqrt{\Delta}}\sinh\theta & \cosh\theta - \frac{h}{\sqrt{\Delta}}\sinh\theta \end{pmatrix}. \tag{6}$$

For $\Delta = 0$ (linear type),

$$S = S(b, h) = \begin{pmatrix} m + hb & b \\ -h^2 b & m - hb \end{pmatrix}, \tag{7}$$

where $m = \pm 1$.

Any 2×2 non-diagonal regular matrix A is represented by the polar form of

$$A = (\det A)^{1/2} S(\theta, k, h) \text{ or } A = (\det A)^{1/2} S(b, h). \tag{8}$$

■ *Extended velocity composition theorem*

We have a velocity composition theorem from Lorentz transformation

$$v_{13} = -v_{31} = (v_{12} + v_{23}) / \left(1 + v_{12}v_{23}/c^2\right).$$

This is equivalently changed to:

$$(c - v_{12})(c - v_{23})(c - v_{31}) = (c + v_{12})(c + v_{23})(c + v_{31}). \tag{9}$$

This is extended 3 to

$$n : (c - v_{12})(c - v_{23}) \cdots (c - v_{n1}) = (c + v_{12})(c + v_{23}) \cdots (c + v_{n1}).$$

2 Introduction

Think everything in a two-dimensional space-time model.

(1) Put right-hand oblique coordinate systems on both face sides of a plane, and make them coincide with their origins. We define a 2×2 back coordinate transformation matrix B as an inside out transformation, then $\det B < 0$. Turn over the back coordinate transformation

B to derive transformation A which transforms between right-hand systems on the front surface.

$$A = MB = (\det A)^{1/2} S(\theta, k, h),$$

where

$$M = \begin{pmatrix} -1 & 0 \\ 0 & 1 \end{pmatrix}, \det A > 0, \ \det S(\theta, k, h) = 1. \tag{10}$$

The special linear transformation $S(\theta, k, h)$ is given by the polar form. The commutative coefficients k, h are the basis of a plane geometry, because they define the symmetry of a plane as shown on item (6).

(2) When $\det A = 1$ and transformations S have common commutative coefficients k, h, they form transformation group based on the invariant function of Equation (4) and they create the isometric transformation geometry with the metric of norm $\|p\| = r$.

(3) If the back coordinate transformation $B \neq B^{-1}$, then we are able to distinguish which side of a plane we are on. Therefore, the symmetry plane equation is

$$B = B^{-1} \Leftrightarrow B^2 = E, \text{ where } \det B < 0. \tag{11}$$

We obtain an *oblique reflection transformation matrix* B with two degrees of freedom:

$$B = \begin{pmatrix} -a & -b \\ c & a \end{pmatrix} = \begin{pmatrix} -a & -b \\ kb & a \end{pmatrix},$$

$$\det B = -1, \ k = c/b, \text{ eigen values } \lambda = \pm 1. \tag{12}$$

(4) Turn over the oblique reflection transformation B to derive transformation F which transforms between right-hand coordinate systems on the front surface,

$$F = MB = \begin{pmatrix} a & b \\ kb & a \end{pmatrix} = S(\theta, k, 0),$$

$$\text{where } M = \begin{pmatrix} -1 & 0 \\ 0 & 1 \end{pmatrix}, \ \det F = 1, \ k = c/b, h = 0, \tag{13}$$

eigen values $\lambda = a \pm \sqrt{k}b$, eigen lines $y = \pm\sqrt{k}x, k, h$ are commutative coefficients.

The matrix $F = S(\theta, k, 0)$ is called a special *iso-diagonal transformation*, and classified by k as follows.

If $k = -1$, then $\boldsymbol{F}$ is referred to as rotation transformation.

If $k = 0$, then $\boldsymbol{F}$ is referred to as Galilean transformation.

If $k > 0$, then $\boldsymbol{F}$ is referred to as Lorentz transformation.

Thus, we obtain these transformations from symmetry of a plane in spacetime. From Equation (4) and $h = 0$, we have the quadratic invariant function $\phi(p)$ of transformation $\boldsymbol{F}$ which satisfies the principle of relativity,

$$\phi(\boldsymbol{F}p) = \phi(p) = -kx^2 + y^2 = r^2, \underline{\det \boldsymbol{F} = 1}. \tag{14}$$

(5) Symmetry axioms: There are five kinds of symmetry of a plane in spacetime.

a. linearity of a plane in spacetime (homogeneity of spacetime) This symmetry includes reversal of space and time, and translation of space and time.

b. commutativity of products of linear transformation

c. front-back symmetry of a plane in spacetime

Next two items are symmetries of a line on a plane [7].

d. isotropy of space by two space lines which are equivalently inverted on a line

e. unidirectional of time by two time lines having same direction

(6) The commutative coefficients k, h define the symmetry of a plane in spacetime.

symt.	h	k	plane	transformation	created geometry
a	-	-	affine(one sided) pln.	$\boldsymbol{A}$ general linear transf.	affine geometry
ab	h	k	front-back asym. pln.	$\boldsymbol{S}(\theta, k, h)$ special linear	asymt. pln. geom.
abc	0	k	front-back symt. pln.	$\boldsymbol{F} = \boldsymbol{S}(\theta, k, 0)$ iso-diagonal	symt. pln. geom.
abcd	0	-1	Euclidean plane	$\boldsymbol{R} = \boldsymbol{S}(\theta, -1, 0)$ rotation	Euclidean geometry
abcde	0	0	Newtonian plane	$\boldsymbol{G} = \boldsymbol{S}(b, 0)$ Galilean	Newtonian mechanics
abcde	0	+	Minkowski plane	$\boldsymbol{L} = \boldsymbol{S}(\theta, k, 0)$ Lorentz	relativity principle

(7) Two inertial coordinate systems moving at a constant speed v on a straight line correspond to the right-hand coordinate systems on the front and back Minkowski plane. This is shown in Figure 1 as a conventional form, and in Figure 2 as a symmetry form and both of them shows the same transformation $\boldsymbol{B}$.

(8) Minkowski plane means that this plane is ruled by Lorentz transformation and holds Minkowski plane geometry with theorems which are formally the same as Euclidean theorems [6]. Euclidean theorems are covariant with respect to rotational transformations because of Equation (14), similar should be the Minkowski plane geometry.

(9) The principle of relativity is not a principle but a property of linear transformation reflecting the basic symmetries of space and time

from which this principle is mathematically derived from eigen plane made of two eigen lines belonging to the oblique reflection transformation $\boldsymbol{B}$.

3   Proof of the special principle of relativity

3.1   Symmetry plane [2]

Put right-hand oblique coordinate systems on both face sides of a plane, and make them coincide with their origins. We define a 2×2 back coordinate transformation matrix $\boldsymbol{B}$ $\left(\left(\begin{array}{c} x_2 \\ y_2 \end{array} \right) = \boldsymbol{B} \left(\begin{array}{c} x_1 \\ y_1 \end{array} \right) \right)$ as an inside out transformation, then $\det \boldsymbol{B} < 0$. Because it is not distinctive which side of a plane is the back or front, the symmetry plane equation is given as per Equation (11), and the solution is an oblique reflection transformation $\boldsymbol{B}$ given as per Equation (12) as shown below again.

$$\boldsymbol{B} = \boldsymbol{B}^{-1} \Leftrightarrow \boldsymbol{B}^2 = \boldsymbol{E}, \text{ where } \det \boldsymbol{B} < 0. \tag{11}$$

$$\boldsymbol{B} = \left(\begin{array}{cc} -a & -b \\ c & a \end{array} \right) = \left(\begin{array}{cc} -a & -b \\ kb & a \end{array} \right), \tag{12}$$

$$\det \boldsymbol{B} = -1, \ k = c/b, \text{ eigen values } \lambda = \pm 1.$$

The signs of variables a, b, and c can be set arbitrarily. Note that it is different from the setting in Eq. (1).

The oblique reflection transformation $\boldsymbol{B}$ has the following properties.

■ An oblique reflection transformation $\boldsymbol{B}$ guarantees that the front and back coordinate systems are congruent, because their equations of coordinate axes have the same form.

Front: y_2-axis: $x_2 = -ax_1 - by_1 = 0, x_2$-axis: $y_2 = kbx_1 + ay_1 = 0,$
Back: y_1-axis: $x_1 = -ax_2 - by_2 = 0, x_1$-axis: $y_1 = kbx_2 + ay_2 = 0.$
(15)

■ Since $\boldsymbol{B}$ has an eigen value of 1, then $\boldsymbol{B}$ has an invariant line $\mathrm{f}(p)$ from Equation (1) with putting $a \to -a$, .

$$\mathrm{f}(\boldsymbol{Bp}) \equiv \mathrm{f}(\boldsymbol{p}) = cx + (a+1)y. \tag{16}$$

■ For $\lambda = 1 \Leftrightarrow$ a fixed-point equation $\boldsymbol{Bp} = \boldsymbol{p}$, this eigen line is called a fold line f:

$$cx + (a-1)y = 0. \tag{17}$$

---

[2] Gray fonts or lines are on the back coordinate plane.

- For $\lambda = -1 \Leftrightarrow$ an inversion equation $\boldsymbol{Bp} = -\boldsymbol{p}$, this eigen line is called an isotropic line g:

$$cx + (a+1)y = 0 \tag{18}$$

This line g is isotropic with regard to the origin and is parallel to an invariant line $f(\boldsymbol{p})$ given as per Equation (16).
- When a point $\boldsymbol{p}$ is in an invariant line $f(\boldsymbol{p}) = $ s, and the point r is the intersection point of a fold line f and an invariant line $f(\boldsymbol{p})$, then in the fold line f, $\boldsymbol{Br} = \boldsymbol{r}$, and in the invariant line, $f(\boldsymbol{Bp}) = f(\boldsymbol{p}) = f(\boldsymbol{r})$ hold. Obtained by translating the vector $(\boldsymbol{p}-\boldsymbol{r})$ onto the isotropic line g,

$$\boldsymbol{B}(\boldsymbol{p}-\boldsymbol{r}) = -(\boldsymbol{p}-\boldsymbol{r}) \Leftrightarrow \boldsymbol{Bp} - \boldsymbol{r} = -\boldsymbol{p} + \boldsymbol{r} \Leftrightarrow \boldsymbol{Bp} + \boldsymbol{p} = 2\boldsymbol{r},$$
$$\text{where } f(\boldsymbol{Bp}) = f(\boldsymbol{p}) = f(\boldsymbol{r}). \tag{19}$$

Since the fixed point r is the midpoint between the point $\boldsymbol{p}$ and $\boldsymbol{Bp}$, the invariant line $f(\boldsymbol{p})$ is isotropic around the fixed point $\boldsymbol{r}$. For the fold line f and the isotropic line g, the product of both slopes is

$$\frac{-c}{a-1}\frac{-c}{a+1} = \frac{c^2}{a^2-1} = \frac{c^2}{bc} = \frac{c}{b} = k. \text{ (orthogonal)} \tag{20}$$

Eigen plane is composed of a fold line f and an isotropic line g, and they are generally oblique in figure but orthogonal in equation. ($\rightarrow$ See Figure 1)

The eigen plane belonging to the oblique reflection transformation B is called the oblique *reflection plane*. As the fixed point r is the midpoint between the point $\boldsymbol{p}$ and $\boldsymbol{Bp}$, this oblique reflection plane has oblique reflection symmetry centered a fold line on the front side, and it is semi-isotropic. It also confirms the symmetry of the front and back coordinate plane from Equation (15). ($\rightarrow$ See Figure 1 and 2)
- Turn over the oblique reflection transformation B to derive transformation F which transforms between right-hand coordinate systems on the front surface as given by Equation (13). This transformation is x-reversal ($x_F = -x_2, y_F = y_2$) of the back coordinate system $x_2 - y_2$ by the reflection transformation M to form right-hand system $x_F - y_F$ on the front side.
- Figure 1. shows the case of $B = \frac{1}{3}\begin{pmatrix} -5 & 4 \\ -4 & 5 \end{pmatrix} = B^{-1}, k = 1$ on a Minkowski plane and $p = \begin{pmatrix} -3 \\ -3 \end{pmatrix}$ for example. Coordinate $(x, y) = (x_1, y_1)$ and (x_F, y_F) are on the front side, coordinate (x_2, y_2) is on the back side.

- The back coordinate system $(x_2 - y_2)$ and the front coordinate system $(x - y)$ are exactly equivalent as shown Equation (15). The fold line f is the midline between y and y_2-axes.
- Transformation B has duality of items a and b.

a. Figure inversion transformation:

$$Bp = q = \begin{pmatrix} 1 \\ -1 \end{pmatrix} = \begin{pmatrix} x \\ y \end{pmatrix} \text{ on the same surface.}$$

b. Back coordinate transformation:

$$Bp = q = \begin{pmatrix} 1 \\ -1 \end{pmatrix} = \begin{pmatrix} x_2 \\ y_2 \end{pmatrix} \text{ from front to back surface.}$$

$$p = \begin{pmatrix} -3 \\ -3 \end{pmatrix}$$

$$p = Bq = \begin{pmatrix} -3 \\ -3 \end{pmatrix}, \quad p = Bq = \begin{pmatrix} -3 \\ -3 \end{pmatrix},$$

$$r = \begin{pmatrix} -1 \\ -2 \end{pmatrix} = (p+q)/2 = \begin{pmatrix} -1 \\ -2 \end{pmatrix} = r$$

$$Bp = q = \begin{pmatrix} 1 \\ -1 \end{pmatrix}, \quad Bp = q = \begin{pmatrix} 1 \\ -1 \end{pmatrix}$$

$$Bq = p = \begin{pmatrix} -3 \\ -3 \end{pmatrix}, \quad Bq = p = \begin{pmatrix} -3 \\ -3 \end{pmatrix}$$

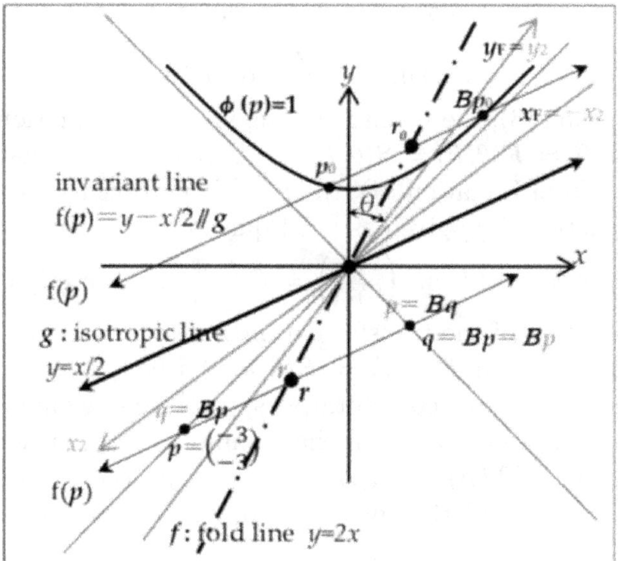

Fig. 1. Oblique reflection plane

Polar form of the special iso-diagonal transformation F

▪ From the polar form of the special linear transformation S, we obtain the polar form of the transformation $F = S(\theta, k, 0), h = 0$. The θ is the angle of intersection which is called a declination created by both y-axis and y_2-axis on the coordinate system [6].

When $k < 0$,

$$F = \begin{pmatrix} a & b \\ kb & a \end{pmatrix} = \begin{pmatrix} \cos\theta & \sin\theta/\sqrt{-k} \\ -\sqrt{-k}\sin\theta & \cos\theta \end{pmatrix}, \qquad (21)$$

where θ is the elliptic angle.

This matrix F is called an elliptic transformation. In particular, when $k = -1$, the matrix F is called a rotational transformation, the matrix $B = MF$ is called a reflection transformation, and they are called orthogonal transformations.

When $k > 0$,

$$F = \begin{pmatrix} a & b \\ kb & a \end{pmatrix} = \begin{pmatrix} \cosh\theta & \sinh\theta/\sqrt{k} \\ \sqrt{k}\sinh\theta & \cosh\theta \end{pmatrix}, \qquad (22)$$

where θ is the hyperbolic angle. This matrix F is called Lorentz transformation.

When $k = 0$,

$$F = \begin{pmatrix} a & b \\ kb & a \end{pmatrix} = \begin{pmatrix} a & b \\ 0 & a \end{pmatrix}, \quad a = \pm 1. \qquad (23)$$

This matrix F is called Galilean transformation.

▪ From Equation (3), the special iso-diagonal transformation F is given by $h = 0$ on $S \Leftrightarrow F(\theta, k) = S(\theta, k, 0)$, because the diagonal element of F is $a = d$. From Equation (4), the transformation F has a quadratic invariant function $\phi(p)$ given as per Equation (14). Since the function $\phi(p)$ is also obtained from $\begin{pmatrix} u \\ v \end{pmatrix} = F \begin{pmatrix} x \\ y \end{pmatrix}$, then $\phi(p)$ is the only invariant function of the transformation F.

▪ The transformation $F(\theta, k) = S(\theta, k, 0)$ rule the symmetry plane transformation, create a commutative special iso-diagonal transformation continuous group based on the commutative coefficient k. Since $B = MF$ and $\phi(Mp) = \phi(p) = -kx^2 + y^2$, the oblique reflection transformation B and the special iso-diagonal transformation F, F^2

and F^{-1} have the common invariant function $\phi(p)$ as shown below.

$$\phi(Bp) = \phi(M(Fp)) = \phi(Fp) = \phi(p) = -kx^2 + y^2,$$

$$\text{where } M = \begin{pmatrix} -1 & 0 \\ 0 & 1 \end{pmatrix}, \ p = \begin{pmatrix} x \\ y \end{pmatrix}, \tag{24}$$

$$\phi\left(F^2 p\right) = \phi(F(Fp)) = \phi(Fp) = \phi(p)$$

$$\phi(p) = \phi\left(F\left(F^{-1}p\right)\right) = \phi\left(F^{-1}p\right)$$

When the commutative coefficient k is fixed to the plane, the matrices F and B have one degree of freedom of θ, and any product of these matrices F and B is closed on the orbit of a common invariant function $\phi(p)$. For example,

$$\phi\left(BF^{2\cdots}B^{-1}F^{-1}p\right) = \phi\left(F^2\cdots B^{-1}F^{-1}p\right) = \phi\left(F\cdots B^{-1}F^{-1}p\right)$$
$$= \phi\left(B^{-1}F^{-1}p\right) = \phi\left(BF^{-1}p\right) = \phi\left(F^{-1}p\right) = \phi(p) = -kx^2 + y^2. \tag{25}$$

Therefore these transformations B and F create an isometric continuous transformation group based on the quadratic invariant function $\phi(p)$.

■ The oblique reflection transformation (back coordinate transformation) B transforms the point p on the front surface to the corresponding point q on the back surface such as

$$q_1 = Bp_1, \ q_2 = Bp_2. \tag{26}$$

On the front side, the figure transformation X with right-hand system transforms the point p_1 to p_2. Similarly, on the back side, the figure transformation Y transforms the point q_1 to q_2 such as

$$p_2 = Xp_1, \ \det X > 0, \ q_2 = Yq_1, \ \det Y > 0. \tag{27}$$

From these four equations, we obtain

$$q_2 = Yq_1 = \underline{YB}p_1 = Bp_2 = \underline{BX}p_1. \tag{28}$$

Since the point p_1 is arbitrary and $B = B^{-1}$, we obtain the following.

$$YB = BX \Leftrightarrow Y = BXB \Leftrightarrow BY = XB,$$
$$\det Y = \det X > 0, \ \operatorname{tr} Y = \operatorname{tr} X. \tag{29}$$

Thus, the matrices X and Y are similar.

Substituting $B = M$ and $B = MF$ into $YB = BX$ in Equation (29) respectively, and therefore

$$YM = MX, \quad \underline{YMF} = MFX = \underline{MXF}. \tag{30}$$

Comparing the second and third sides, $FX = XF$ and similarly $FY = YF$.

The right-hand coordinate transformation F and the figure transformation X, Y are commutative and have the same commutative coefficient k.

When det $X = 1$, the transformations B, F, X, Y and any product of them have a common quadratic invariant function $\phi(p)$ on both sides of a plane such that

$$\phi(Bp) = \phi(Fp) = \phi(Xp) = \phi(Yp) = \phi(p) = -kx^2 + y^2 = r^2 \tag{31}$$

as an orbit, and the geometry of the isometric transformation continuous group with the norm $\|p\| = r$ is created.

$$\|Bp\| = \|Fp\| = \|Xp\| = \|Yp\| = \|p\| = r. \tag{32}$$

3.2   What k gives? Three types of a symmetry plane

For one oblique reflection transformation B, there corresponds one special iso-diagonal transformation F and one oblique reflection plane. The declination θ is the angle of intersection between y_1-axis and y_2-axis created by the transformation B. When the declination θ is a whole real number, the existing region of the fold line f and the isotropic line g depends on the positive or negative commutative coefficient k.

(1) When $k < 0, F$ is elliptic transformation. The polar form of the transformation $B = MF$ is from Equation (21),

$$B = \begin{pmatrix} -a & -b \\ c & a \end{pmatrix} = \begin{pmatrix} -\cos\theta & -\sin\theta/\sqrt{-k} \\ -\sqrt{-k}\sin\theta & \cos\theta \end{pmatrix},$$
$$b = \frac{1}{\sqrt{-k}}\sin\theta, \quad c = kb. \tag{33}$$

The fold line f and the isotropic line g are obtained from Equations

(17) and (18).

the fold line f :

$$y = \frac{-c}{a-1}x = \sqrt{-k}\frac{\sin\theta}{\cos\theta - 1}x = \sqrt{-k}\cot\frac{\theta}{2} \cdot x = ux,$$

the isotropic line g : $\qquad\qquad(34)$

$$y = \frac{-c}{a+1}x = \sqrt{-k}\frac{\sin\theta}{\cos\theta + 1}x = \sqrt{-k}\tan\frac{\theta}{2} \cdot x = vx,$$

the slopes of eigen lines $f, g : -\infty < u < \infty, \infty < v < \infty$.

The half angle of declination θ determines the azimuth of the fold and isotropic line. When the elliptic angle $\theta = \angle y_1 y_2$ is a whole real number, the existence of the fold line representing the oblique reflection plane is evenly distributed and completely isotropic around the origin. Therefore, this symmetry plane is thought to be space $\times$ space type plane. This plane is called the extended Euclidean plane or elliptic type plane, and extended Euclidean plane geometry holds. In particular, when $k = -1$, the invariant function $\phi(p) = x^2 + y^2$ is a circle and this plane holds Euclidean geometry.

(2) When $k > 0, F$ is Lorentz transformation. The polar form of the transformation $B = MF$ is from Equation (22),

$$B = \begin{pmatrix} -a & -b \\ c & a \end{pmatrix} = \begin{pmatrix} -\cosh\theta & -\sinh\theta/\sqrt{k} \\ \sqrt{k}\sinh\theta & \cosh\theta \end{pmatrix},$$

$$b = \frac{1}{\sqrt{k}}\sinh\theta, c = kb. \qquad\qquad(35)$$

The fold line f and the isotropic line g are obtained from Equations (17) and (18).

the fold line f :

$$y = \frac{-c}{a-1}x = -\sqrt{k}\frac{\sinh\theta}{\cosh\theta - 1}x = -\sqrt{k}\coth\frac{\theta}{2} \cdot x = ux,$$

the isotropic line g : $\qquad\qquad(36)$

$$y = \frac{-c}{a+1}x = -\sqrt{k}\frac{\sinh\theta}{\cosh\theta + 1}x = -\sqrt{k}\tanh\frac{\theta}{2} \cdot x = vx.$$

the asymptotic line $: y = \pm\sqrt{k}x. \qquad\qquad(37)$

The half angle of declination θ determines the azimuth of the fold and isotropic line. The regions of existence of the eigen lines f and g are :

195

The slope u of the fold line f is $-\infty < u < -\sqrt{k}$ and $\sqrt{k} < u < \infty$ in the upper and lower quadrants partitioned by both asymptotes. The slope v of the isotropic line g is $-\sqrt{k} < v < \sqrt{k}$ in the left and right quadrants. $\rightarrow$ See Fig. 1.

When the hyperbolic angle $\theta = \angle y_1 y_2$ is a whole real number, the existence of the fold line representing the oblique reflection plane is unevenly distributed in the upper and lower quadrants. A point p on the invariant function $\phi(p)$, namely upward hyperbola, is transitively transformed across the fold line f to a point Bp on the same hyperbola $\phi(Bp) = \phi(p)$ in the same quadrant, so both y-coordinates of p and Bp are greater than 0. Since the azimuths of all the fold lines are aligned to upward and all the isotropic lines $p - Bp$ are isotropic, then this symmetry plane is thought to be space $\times$ time type plane. This semi-isotropic plane is called Minkowski plane, and holds hyperbolic type plane geometry, i.e, Minkowski plane geometry.

(3) When $k = 0, F$ is the Galilean transformation. From Equation (23), we obtain

$$B = MF = \begin{pmatrix} -a & -b \\ 0 & a \end{pmatrix}, a = \pm 1, \det B = -1.$$

the fold line$f : x = -by/2$ (when $a = 1$),
the isotropic line $g : y = 0$ (x-axis), $\qquad\qquad$ (38)
the invariant line : f(Bp) = f(p) = y,

the invariant function: $\phi(Bp) = \phi(Fp) = \phi(p) = y^2$.

When we think $y = t$ (time), time is an invariant quantity from the invariant function $\phi(F\ p) = \phi(p) = t^2$, which is consistent with Newton's idea of absolute time, so this plane is called a Newtonian plane. This plane is semi-isotropic, and the spatial axis ($y = 0$) of the oblique reflection planes is shared.

From the above, three types of the symmetry plane are extended Euclidean plane and Minkowski plane and Newtonian plane. The geometries of these planes satisfy the extended Curie's principle, and also the principle of relativity from Equation (14). Thus, the laws on the symmetry plane must be made from the invariant function $\phi(p)$ with the norm $\|p\| = r$, because they must be the special iso-diagonal transformation F invariant.

[End of proof]

3.3 Symmetric spacetime structure of two inertial coordinate systems

Consider two inertial coordinate systems S1 and S2 moving away from each other on a line with constant velocity v. Both systems are equivalent. When each of the two inertial systems sees the other from a space × time right-hand system, the velocity and the positive orientation of the space axis are reversely related to each other, so the two systems correspond to the two sides of the Minkowski plane. The origins of both inertial coordinate systems are made to coincide. In the Figure 2, the coordinate axis $x_1 - t_1$ of $S1$ is on the front side and $x_2 - t_2$ of S2 is on the back side. The back coordinate transformation is the oblique reflection transformation $\mathbf{B}$. On the Minkowski plane, the special iso-diagonal transformation $\mathbf{F}$ is Lorentz transformation $\mathbf{L}(=\mathbf{MB})$ which transforms from coordinate system S1 to Lorentz coordinate system SL such that

$$\begin{pmatrix} x_2 \\ t_2 \end{pmatrix} = \mathbf{B} \begin{pmatrix} x_1 \\ t_1 \end{pmatrix} \text{ where } \mathbf{B} = \begin{pmatrix} -a & -b \\ kb & a \end{pmatrix},$$

multiplying both sides by $\mathbf{M} = \begin{pmatrix} -1 & 0 \\ 0 & 1 \end{pmatrix}$,

$$\mathbf{M} \begin{pmatrix} x_2 \\ t_2 \end{pmatrix} = \mathbf{MB} \begin{pmatrix} x_1 \\ t_1 \end{pmatrix}, \tag{39}$$

therefore $\begin{pmatrix} -x_2 \\ t_2 \end{pmatrix} = \begin{pmatrix} x_L \\ t_L \end{pmatrix} = L \begin{pmatrix} x_1 \\ t_1 \end{pmatrix},$

where $\mathbf{L} = \mathbf{MB} = \begin{pmatrix} a & b \\ kb & a \end{pmatrix}$.

The expansion formula for the Lorentz transformation is

$$x_L = ax_1 + bt_1, \ t_L = kbx_1 + at_1, \text{ where } k > 0.$$

From the first equation, a is a unitless constant and b is a velocity constant. In the $x_1 - t_1$ coordinate system, the motion of $S2$ is expressed as $x_1 = vt_1$, while the expression for the t_L axis is $x_L = ax_1 + bt_1 = 0$, so $v = x_1/t_1 = -b/a$. In the second equation, k is the inverse of the velocity squared, and we use the velocity constant c as $k = 1/c^2$ by convention. As $\det \mathbf{L} = 1$, so we obtain $a = \left(1 - v^2/c^2\right)^{-1/2} = \gamma \geqq 1$. Thus, from the two inertial coordinate systems with velocity v, the Lorentz transformation $\mathbf{L}$ and its oblique reflection transformation $\mathbf{B}$

are specifically obtained as follows.

$$L = \gamma \begin{pmatrix} 1 & -v \\ -v/c^2 & 1 \end{pmatrix}, \; B = \gamma \begin{pmatrix} -1 & v \\ -v/c^2 & 1 \end{pmatrix} = ML,$$

$$\det B = -1, \; L \begin{pmatrix} 0 \\ 1 \end{pmatrix} = \gamma \begin{pmatrix} -v \\ 1 \end{pmatrix}$$

$$L^{-1} = \gamma \begin{pmatrix} 1 & v \\ v/c^2 & 1 \end{pmatrix}, \; L^{-1} \begin{pmatrix} x_L \\ t_L \end{pmatrix} = \begin{pmatrix} x_1 \\ t_1 \end{pmatrix},$$

$\gamma = 1/\sqrt{1 - v^2/c^2}$, c is a velocity constant.

Figure 2 shows the same $B = \frac{1}{3} \begin{pmatrix} -5 & 4 \\ -4 & 5 \end{pmatrix}$ as Figure 1 with $k = c = 1, v = 4/5, \gamma = 5/3$.

Inertial frame 1: S1 (x_1, t_1),
Inertial frame $2 : S2\,(x_2, t_2)$,
Lorentz coordinate system: SL (x_L, t_L)

Redraw Figure 1 with the fold line and isotropic line as cross centerlines to obtain Figure 2, but replace y axis with t axis.

Fold line $f : t = 2x$,
isotropic line $g : t = x/2$.

$$L \begin{pmatrix} v \\ 1 \end{pmatrix} = \begin{pmatrix} 0 \\ 1/\gamma \end{pmatrix},$$

$$L \begin{pmatrix} \gamma v \\ \gamma \end{pmatrix} = \begin{pmatrix} 0 \\ 1 \end{pmatrix},$$

$$L \begin{pmatrix} 1 \\ 0 \end{pmatrix} = \gamma \begin{pmatrix} 1 \\ -v/c^2 \end{pmatrix},$$

$$B \begin{pmatrix} v \\ 1 \end{pmatrix} = \begin{pmatrix} 0 \\ 1/\gamma \end{pmatrix},$$

$$f : B \begin{pmatrix} 3 \\ 6 \end{pmatrix} = \begin{pmatrix} 3 \\ 6 \end{pmatrix},$$

$$g : B \begin{pmatrix} 6 \\ 3 \end{pmatrix} = \begin{pmatrix} -6 \\ -3 \end{pmatrix},$$

$$x = ct : B \begin{pmatrix} 1 \\ 1 \end{pmatrix} = \begin{pmatrix} -1/3 \\ 1/3 \end{pmatrix},$$

$$L \begin{pmatrix} 1 \\ 1 \end{pmatrix} = \begin{pmatrix} 1/3 \\ 1/3 \end{pmatrix}.$$

(40)

198

Any two inertial coordinate systems on a line are drawn as symmetrical as in Figure 2.

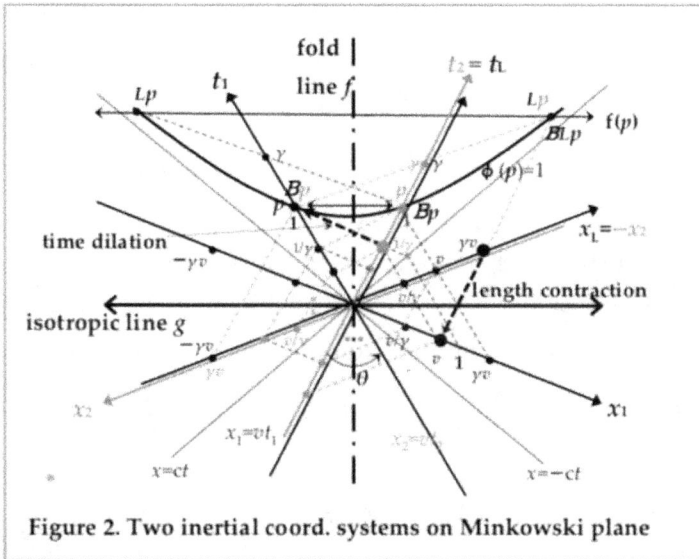

Figure 2. Two inertial coord. systems on Minkowski plane

Maximum universal speed c

From Equation (9), we obtain the limit speed of inertial frames, where c is a velocity constant.

$$(c - v_{12})(c - v_{23}) \cdots (c - v_{n1}) = (c + v_{12})(c + v_{23}) \cdots (c + v_{n1})$$
$$\Leftrightarrow (c - v_{12})^2 (c - v_{23})^2 \cdots (c - v_{n1})^2 = \tag{9}$$
$$= (c^2 - v_{12^2})(c^2 - v_{23^2}) \cdots (c^2 - v_{n1}^2) \geqq 0,$$

and n is integer $\Leftrightarrow c^2 - v_{ij}^2 \leqq 0 \Leftrightarrow (v_{ij} - c)(v_{ij} + c) \leqq 0,$

therefore $-c \leqq v_{ij} \leqq c,$ $\tag{41}$

where v_{ij} is the speed of inertial frame from i to j.

Any speed of inertial frame v_{ij} is not able to exceed the velocity constant c which means the maximum universal speed. From Maxwell's law, the speed of light in the inertial reference frame is constant and the maximum speed in nature is the light speed in vacuum. On the basis of the Minkowski plane $(h = 0, k = 1/c^2)$ that guarantees the front-back symmetry in the empty spacetime, the maximum universal velocity constant c must be the speed of light in vacuum. This is equivalent to applying the principle of relativity to Maxwell's law of the speed of light.

4   Conclusion

It is noted that any function f(p) held on the front side holds the same as function f(p) on the back side such as

Back coordinate transformation $\boldsymbol{B}\boldsymbol{p} = q$, figure inversion transformation $\boldsymbol{B}q = p$,

$$\boldsymbol{B} = \boldsymbol{B}^{-1}, \boldsymbol{p} = \boldsymbol{B}^{-1}q = \boldsymbol{B}q = p, \text{ therefore f}(\boldsymbol{p}) = \text{f}(p). \tag{42}$$

The transformation $\boldsymbol{B}$ plays the role of intermediary in the front-back symmetry. On the other hand, the transformation $\boldsymbol{F}$ satisfies the principle of relativity, because it keeps the same function of $\phi(\boldsymbol{p}) = \phi(\boldsymbol{q})$, $\boldsymbol{q} = \boldsymbol{F}\boldsymbol{p}$ on the different right-hand coordinate systems of $\boldsymbol{p} = \begin{pmatrix} x \\ y \end{pmatrix}$ and $\boldsymbol{q} = \begin{pmatrix} u \\ v \end{pmatrix}$ given as per Equation (14),

$$\underline{\phi(\boldsymbol{F}\boldsymbol{p}) = \phi(\boldsymbol{p})} = -kx^2 + y^2 = r^2, \underline{\det \boldsymbol{F} = 1}. \tag{14}$$

Also this transformation $\boldsymbol{F}$ satisfies the extended Curie's principle as shown in section 3.2. Thus, the only invariant function $\phi(\boldsymbol{p})$ of transformation $\boldsymbol{F}$ should be the core of the law of a symmetry plane. Furthermore, to be transformation $\boldsymbol{F}$ invariant, the law of spacetime must be a variation of invariant function $\phi(\boldsymbol{p})$ with its invariant r such that

$$\phi\left(\boldsymbol{F}\left(\boldsymbol{p}_1 \pm \boldsymbol{p}_2\right)\right) = \phi\left(\boldsymbol{p}_1 \pm \boldsymbol{p}_2\right) \quad \text{or} \quad \phi\left(\boldsymbol{F}\frac{d}{dr}\boldsymbol{p}\right) = \phi\left(\frac{d}{dr}\boldsymbol{p}\right). \tag{43}$$

We are able to find these variations in the basic laws of physics on Minkowski plane or Newtonian plane and theorems of geometry on Euclidean plane or Minkowski plane based on Equations (31) and (32) [6]. For example, the inner product of the symmetry plane is derived from $\phi(\boldsymbol{F}\boldsymbol{p}) = \phi(\boldsymbol{p}) = -kx^2 + y^2 = r^2 = \|\boldsymbol{p}\|^2$ such that

$$\left(\phi\left(\boldsymbol{F}\boldsymbol{p}_1\right) + \phi\left(\boldsymbol{F}\boldsymbol{p}_2\right) - \phi\left(\boldsymbol{F}\left(\boldsymbol{p}_1 - \boldsymbol{p}_2\right)\right)\right)/2$$
$$= \left(\phi\left(\boldsymbol{p}_1\right) + \phi\left(\boldsymbol{p}_2\right) - \phi\left(\boldsymbol{p}_1 - \boldsymbol{p}_2\right)\right)/2$$
$$= -kx_1x_2 + y_1y_2 = \left(\boldsymbol{p}_1, \boldsymbol{p}_2\right).$$

My paper establishes the followings. On the basis of the structure of a linear plane, from the symmetry of the space $\times$ space type plane (Euclidean plane), we have the reflection and the rotation transformation group which form Euclidean geometry. Also, from the symmetry of the space $\times$ absolute time type plane (Newtonian plane), we have the oblique reflection and the Galilean transformation group which form Newtonian mechanics.

Finally, from the symmetry of the space × time type plane (Minkowski plane), we obtain the oblique reflection and the Lorentz transformation group which form the principle of relativity. Therefrom, Minkowski plane geometry and the special theory of relativity are established [2][3].

Acknowledgements

I would like to thank Dr. Gregorie Dupuis-Mc Donald for his philosophical and sharp advice in the preparation of this paper.

Reference

[1] A. Einstein, "On the electrodynamics of moving bodies" 1905

[2] H. Poincaré, *"Électricité et optique"* 1901

[3] H. Poincaré, *Science and Hypothesis* 1902 "The reason why true proof produces various results is that the conclusion is in a sense more general than the premise."

[4] Hiroaki Fujimori, web site http://www.spatim.sakura.ne.jp/

[5] Hiroaki Fujimori, YouTube "Relativity Arises from the Symmetry of Spacetime" via website [4] − tag [Video]

[6] Hiroaki Fujimori, "表裏対称平面の幾何 *Geometry of Symmetry plane*" Bun-shin Press 2022

[7] Hiroaki Fujimori, *Symmetry plane causes relativity and rotation* 2021 http://www.spatim.sakura.ne.jp/pdfpp/sym_pln.pdf/

10 NONCOMPACTIFIED KALUZA-KLEIN THEORIES

S. M. M. RASOULI

Abstract   We provide an overview of noncompactified Kaluza–Klein theories. The space–time-matter theory (or induced-matter theory) and the modified Brans–Dicke theory are discussed. Finally, an extended version of the Kantowski–Sachs anisotropic model is investigated as a cosmological application of the latter.

Keywords: Kaluza–Klein theories; modified Brans–Dicke theory; induced–matter theory; modified Sáez–Ballester theory; noncompactified cosmology; Kantowski-Sachs model

1 Introduction

An outline of the original Kaluza–Klein (KK) theory is given in this section. Subsequently, in a few succinct sentences, we concentrate on the noncompactified KK theories, their motives, and how they differ from their compactified counterparts. For a more in-depth analysis, see [1].

The close relationship between Minkowski's four-dimensional (4D) spacetime and Maxwell's unification of electricity and magnetism may have been the main inspiration for Nördstrom and Kaluza, who were the first physicists to attempt to unify gravity and electromagnetism using a 5D framework. Both have assumed that all derivatives with respect to the fifth coordinate $x^4 \equiv l$ must be zero (cylinder condition) [1].

Let us give a brief overview of a generalized version of the Kaluza mechanism, for a detailed study see [1, 2]. (i) Einstein's field equations were assumed in five dimensions along with the Kaluza's first key assumption: *"the universe in higher dimensions is empty."* [1]. (ii) The

A. S. Stefanov, G. Dupuis-Mc Donald (Eds), *Spacetime Conference - 2022.*
Selected peer-reviewed papers presented at the Sixth International Conference on the Nature and Ontology of Spacetime, 12 - 15 September 2022, Albena, Bulgaria (Minkowski Institute Press, Montreal 2023). ISBN 978-1-989970-96-6 (softcover), ISBN 978-1-989970-97-3 (ebook).

definitions of the Christoffel symbols and the 5D Ricci tensor are taken exactly as they are defined in four dimensions (*Kaluza's second key assumption*). (iii) The cylinder condition has been assumed as the *third key assumption*. By combining the aforementioned assumptions with the specific 5D line–element $\mathcal{G}_{ab}^{\text{KK}}$ (Throughout this review paper, we take the signature of the 4D metric as (- + + +). Moreover, the small Latin indices and the Greek indices run over $0, 1, 2, 3, 4$ and $0, 1, 2, 3$, respectively.) with components

$$\mathcal{G}_{\mu\nu}^{\text{KK}} = g_{\mu\nu} + \psi^2 A_\mu A_\nu, \quad \mathcal{G}_{5\nu}^{\text{KK}} = \mathcal{G}_{\nu 5}^{\text{KK}} = \psi^2 A_\nu, \quad \mathcal{G}_{55}^{\text{KK}} = \psi^2, \quad (1.1)$$

(Where $g_{\mu\nu}$, A_μ and ψ denote the 4D metric tensor, electromagnetic potential and a scalar field, respectively.) 15 field equations are obtained on a 4D hypersurface. With KK assumption, $\psi = 1$, one obtains the Einstein equations (with the electromagnetic energy-momentum tensor) and Maxwell equations on the hypersurface.

The fifth dimension, as mentioned in Kaluza's mechanism, exists, but according to Kaluza's first key assumption, the derivatives of physical quantities with respect to the fifth coordinate l should vanish. However, in the *compactified KK theory*, the fifth coordinate was assumed to be a lengthlike with two properties: a circular topology and a small scale, in accordance with *Klein's assumption*. Given the former, any quantity can become periodic, so the only modes that are independent of the fifth coordinate are observables, as required by [1].

In the noncompactified KK theories, which are the main focus of this review, the idea that the new coordinates are physical still remains, but the compactified approach is generalized by omitting the Kaluza's third key assumption. This means that all the physical quantities can depend on the extra coordinate. Therefore, in these noncompactified frameworks, the equations of motion are modified by the dependence on extra coordinate. Consequently, not only the electromagnetic radiation emerge from geometry, but also a very general kind of matter. It is worth noting that in noncompactified theories the extra coordinates should not necessarily be assumed to be lengthlike.

One of the most well-known noncompactified KK theories is the space-time-matter theory or the induced-matter theory (IMT) [3, 4], which is briefly introduced in the next section. In the IMT, using Einstein's general theory of relativity (GR) as the underlying theory and using an appropriate reduction method, it has been shown that 5D KK equations without sources generates Einstein equations with the induced energy-momentum tensor (EMT). Cosmological/astrophysical applications of the IMT (or its extended versions) are widespread in the literature, see [4, 1, 5], and references therein. For instance, interesting solutions were obtained in Ref. [6] by considering a conformally flat bulk space. Then, the energy conditions associated with this model have also been studied in [7].

Recently, generalized versions of the IMT have been established by applying some alternative theories to GR as the underlying theory: the modified Sáez–Ballester theory (MSBT) [8, 9, 10] and the Modified Brans–Dicke Theory (MBDT) [11, 13] are two important examples. One of the main goals of this review will be the latter and its cosmological application.

This review paper is structured as follows. In the next section we give a brief overview of the IMT. In the Section 3 we describe the field equations of the MBDT. In the 4 section, considering the $5D$ empty Kantowski-Sachs (KS) universe, the BD field equations (we call it BD-KS cosmology) and a new time coordinate, we get the corresponding exact solutions. Then the corresponding constraints among the parameters of the model and some special cases are presented. The Section 5 discusses the properties of physical quantities associated with the BD-KS cosmology of the 4D hypersurface. Finally, in Section 6 we present a summary and further discussion.

2   Five-dimensional Ricci–Flat Space and the IMT

In this section, we present a brief overview of the IMT, for a detailed study see [14] and the references therein.

Considering GR as the underlying theory in an empty 5D spacetime, and applying a reduction method, the field equations of GR with sources can be set up on a 4D hypersurface.

By considering the 5D metric as

$$dS^2 = \mathcal{G}_{ab}(x^c)dx^a dx^b, \tag{2.1}$$

where our 4D universe can be embedded locally and isometrically, is taken as

$$dS^2 = g_{\mu\nu}(x^\alpha, l)dx^\mu dx^\nu + \epsilon\psi^2(x^\alpha, l)\,dl^2. \tag{2.2}$$

In equation (2.2), $\psi = \psi(x^\alpha, l)$ is a scalar field and $\epsilon = \pm 1$ (where $\epsilon^2 = 1$) is an indicator that allows the extra dimension to be selected as time–like or space–like. In order to keep in touch with the content of the original papers in each section, we use the same units that were included therein. For example, in this section we take the same units used in [14]. The metric (2.2) is the KK metric (1.1) with $A_\mu = 0$ (further justification was presented in [1]).

The $\alpha\beta-$, $\alpha 4-$ and $44-$parts of the Ricci tensor $R_{ab}^{(5)}$ are:

$$R_{\alpha\beta}^{(5)} = R_{\alpha\beta}^{(4)} - \frac{\mathcal{D}_\alpha \mathcal{D}_\beta \psi}{\psi} + \frac{\epsilon}{2\psi^2}\left(\frac{\overset{*}{\psi}\overset{*}{g}_{\alpha\beta}}{\psi} - \overset{**}{g}_{\alpha\beta}\right. \tag{2.3}$$
$$\left. + \ g^{\lambda\mu}\overset{*}{g}_{\alpha\lambda}\overset{*}{g}_{\beta\mu} - \frac{1}{2}g^{\mu\nu}\overset{*}{g}_{\mu\nu}\overset{*}{g}_{\alpha\beta}\right),$$

$$R_{4\alpha}^{(5)} = \psi \mathcal{D}_\beta P^\beta{}_\alpha, \tag{2.4}$$

$$R_{44}^{(5)} = -\epsilon\psi\mathcal{D}^2\psi - \frac{1}{4}\overset{*}{g}^{\lambda\beta}\overset{*}{g}_{\lambda\beta} - \frac{1}{2}g^{\lambda\beta}\overset{**}{g}_{\lambda\beta} + \frac{\overset{*}{\psi}}{2\psi}g^{\lambda\beta}\overset{*}{g}_{\lambda\beta}, \tag{2.5}$$

where

$$P_{\alpha\beta} \equiv \frac{1}{2\psi}\left(\overset{*}{g}_{\alpha\beta} - g_{\alpha\beta}g^{\mu\nu}\overset{*}{g}_{\mu\nu}\right), \tag{2.6}$$

$\overset{*}{A} \equiv \frac{\partial A}{\partial l}$, $\mathcal{D}_\alpha$ denotes the covariant derivative on the 4D hypersurface and $\mathcal{D}^2 \equiv \mathcal{D}^\alpha \mathcal{D}_\alpha$.

Assuming a 5D space-time without sources [14, 1, 4] (namely, $G_{ab}^{(5)} = 0 = R_{ab}^{(5)}$, where $R_{ab}^{(5)}$ is the Ricci tensor) and defining a 4D hypersurface Σ_4, where $l = l_0 = $ constant, and

$$g_{\mu\nu}(x^\alpha) = \mathcal{G}^{(4)}(x^\alpha, l_0), \tag{2.7}$$

equations (2.3)–(2.5) yield

$$R_{\alpha\beta}^{(4)} = \frac{\mathcal{D}_\alpha \mathcal{D}_\beta \psi}{\psi} - \frac{\epsilon}{2\psi^2}\left[\frac{\overset{*}{\psi}\overset{*}{g}_{\alpha\beta}}{\psi} - \overset{**}{g}_{\alpha\beta} + g^{\lambda\mu}\overset{*}{g}_{\alpha\lambda}\overset{*}{g}_{\beta\mu}\right. \tag{2.8}$$
$$\left. - \frac{1}{2}g^{\mu\nu}\overset{*}{g}_{\mu\nu}\overset{*}{g}_{\alpha\beta}\right],$$

$$\mathcal{D}_\beta P^\beta{}_\alpha = 0, \tag{2.9}$$

$$\epsilon\psi\mathcal{D}^2\psi = -\frac{1}{4}\overset{*}{g}^{\lambda\beta}\overset{*}{g}_{\lambda\beta} - \frac{1}{2}g^{\lambda\beta}\overset{**}{g}_{\lambda\beta} + \frac{\overset{*}{\psi}}{2\psi}g^{\lambda\beta}\overset{*}{g}_{\lambda\beta}. \tag{2.10}$$

Equations (2.8)–(2.10) " *form the basis of five-dimensional non-compactified KK theory*" [1].

Employing (2.8), (2.10), and $(\delta_\nu^\mu)_{,4} = 0 = g^{\mu\beta} g^{\sigma\lambda} \overset{*}{g}_{\lambda\beta} \overset{*}{g}^{\mu\sigma} + \overset{*}{g}_{\mu\sigma} \overset{*}{g}_{\mu\sigma}$, we get $R^{(4)} = g^{\alpha\beta} R^{(4)}_{\alpha\beta}$ as

$$R^{(4)} = \frac{\epsilon}{4\psi^2} \left[\overset{*}{g}{}^{\mu\nu} \overset{*}{g}_{\mu\nu} + \left(g^{\mu\nu} \overset{*}{g}_{\mu\nu} \right)^2 \right]. \tag{2.11}$$

If we define the EMT in four dimensions as $T^{[\mathrm{IMT}]}_{\alpha\beta} \equiv R^{(4)}_{\alpha\beta} - 1/2 R^{(4)} g_{\alpha\beta}$, then from (2.8) and (2.11), we can easily get

$$T^{[\mathrm{IMT}]}_{\alpha\beta} \equiv \frac{\mathcal{D}_\alpha \mathcal{D}_\beta \psi}{\psi} - \frac{\epsilon}{2\psi^2} \left(\frac{\overset{*}{\psi}\overset{*}{g}_{\alpha\beta}}{\psi} - \overset{**}{g}_{\alpha\beta} + g^{\lambda\mu} \overset{*}{g}_{\alpha\lambda} \overset{*}{g}_{\beta\mu} - \frac{1}{2} g^{\mu\nu} \overset{*}{g}_{\mu\nu} \overset{*}{g}_{\alpha\beta} \right)$$
$$- \frac{\epsilon g_{\alpha\beta}}{8\psi^2} \left[\overset{*}{g}{}^{\mu\nu} \overset{*}{g}_{\mu\nu} + \left(g^{\mu\nu} \overset{*}{g}_{\mu\nu} \right)^2 \right]. \tag{2.12}$$

We can consider $T^{[\mathrm{IMT}]}_{\alpha\beta}$ as an EMT of our 4D universe, known as *induced-matter* in the noncompactified KK theory. Let us be more precise. This induced matter is actually a manifestation of the pure geometry of the higher-dimensional space-time. It is worth noting that the equations on a 4D hypersurface, i.e., $G^{(4)}_{\alpha\beta} = T^{[\mathrm{IMT}]}_{\alpha\beta}$, are automatically included in the corresponding 5D vacuum counterparts $G^{(5)}_{ab} = 0$.

3 Modified Brans-Dicke Theory

In this section we would like to give an overview of the framework established in Ref.[11]. Here, we also consider the metric (2.2) and the noncompact extra dimension. Moreover, equations (2.3)–(2.6) are valid for our herein framework. It should be noted, however, that since the BD scalar field can play the role of a higher–dimensional matter, equations $G^{(5)}_{ab} = 0 = R^{(5)}_{ab} = 0$ are generally no longer valid.

The action of the Brans–Dicke (BD) theory in five dimensions in the Jordan frame, in analogy with its standard counterpart, can be written as

$$\mathcal{S}^{(5)}_{\mathrm{BD}} = \int d^5 x \sqrt{\left| \mathcal{G}^{(5)} \right|} \left[\phi R^{(5)} - \frac{\omega}{\phi} \mathcal{G}^{ab} (\nabla_a \phi)(\nabla_b \phi) + 16\pi L^{(5)}_{\mathrm{matt}} \right], \tag{3.1}$$

where ϕ is the BD scalar field, ω is a dimensionless parameter and ∇_a denotes the covariant derivative in the $5D$ space-time, respectively.

The BD field equations derived from the action (3.1) are:

$$G^{(5)}_{ab} = \frac{8\pi}{\phi} T^{(5)}_{ab} + \frac{\omega}{\phi^2} \left[(\nabla_a \phi)(\nabla_b \phi) - \frac{1}{2} \mathcal{G}_{ab} (\nabla^c \phi)(\nabla_c \phi) \right]$$
$$+ \frac{1}{\phi} \left(\nabla_a \nabla_b \phi - \mathcal{G}_{ab} \nabla^2 \phi \right), \tag{3.2}$$

$$\frac{2\omega}{\phi}\nabla^2\phi - \frac{\omega}{\phi^2}\mathcal{G}^{ab}(\nabla_a\phi)(\nabla_b\phi) + \overset{(5)}{R} = 0, \tag{3.3}$$

where $\nabla^2 \equiv \nabla_a\nabla^a$. On the other hand, from equation (3.2), we obtain

$$\overset{(5)}{R} = -\frac{16\pi\,\overset{(5)}{T}}{3\phi} + \frac{\omega}{\phi^2}(\nabla^c\phi)(\nabla_c\varphi) + \frac{8}{3}\left(\frac{\nabla^2\phi}{\phi}\right). \tag{3.4}$$

where $\overset{(5)}{T} = \mathcal{G}^{ab}\overset{(5)}{T_{ab}}$.

Substituting $\overset{(5)}{R}$ from equation (3.4) into (3.3), we obtain

$$\nabla^2\phi = \frac{8\pi\overset{(5)}{T}}{3\omega + 4}. \tag{3.5}$$

A suitable reduction method was used to establish the MBDT [11], see also [15] for a corrected version of some field equations in the D dimensions. Therefore, let us confine ourselves in this paper to presenting only a summary of a special case where there is no higher–dimensional ordinary matter, i.e., $\overset{(5)}{L_{\text{matt}}} = 0$. Concretely, considering (2.2) as a background metric, equations (3.2) and (3.5) yield four sets of effective field equations on the hypersurface:

1. An equation associated with the scalar field ψ:

$$
\begin{aligned}
\frac{\mathcal{D}^2\psi}{\psi} &= -\frac{(\mathcal{D}_\alpha\psi)(\mathcal{D}^\alpha\phi)}{\psi\phi} \\
&\quad - \frac{\epsilon}{2\psi^2}\left(g^{\lambda\beta}\overset{**}{g}_{\lambda\beta} + \frac{1}{2}\overset{*\lambda\beta}{g}\overset{*}{g}_{\lambda\beta} - \frac{g^{\lambda\beta}\overset{*}{g}_{\lambda\beta}\overset{*}{\psi}}{\psi}\right) \\
&\quad - \frac{\epsilon}{\psi^2\phi}\left[\overset{**}{\phi} + \overset{*}{\phi}\left(\frac{\omega\overset{*}{\phi}}{\phi} - \frac{\overset{*}{\psi}}{\psi}\right)\right].
\end{aligned} \tag{3.6}
$$

2. An effective field equation that is the counterpart of (3.2):

$$
\begin{aligned}
\overset{(4)}{G_{\mu\nu}} &= \frac{8\pi T_{\mu\nu}^{[\text{MBDT}]}}{\phi} + \frac{\omega}{\phi^2}\left[(\mathcal{D}_\mu\phi)(\mathcal{D}_\nu\phi) - \frac{1}{2}g_{\mu\nu}(\mathcal{D}_\alpha\phi)(\mathcal{D}^\alpha\phi)\right] \\
&\quad + \frac{1}{\phi}\left(\mathcal{D}_\mu\mathcal{D}_\nu\phi - g_{\mu\nu}\mathcal{D}^2\phi\right) - g_{\mu\nu}\frac{V(\phi)}{2\phi}. \tag{3.7}
\end{aligned}
$$

where we obtain the induced potential $V(\phi)$ from equation (3.11) (see below); the effective energy–momentum tensor $T_{\mu\nu}^{[\text{MBDT}]}$, in turn, consists of three parts:

$$\frac{8\pi}{\phi}T_{\mu\nu}^{[\text{MBDT}]} \equiv T_{\mu\nu}^{[\text{IMT}]} + \frac{1}{\phi}T_{\mu\nu}^{[\phi]} + \frac{V(\phi)}{2\phi}g_{\mu\nu}. \tag{3.8}$$

Regarding equation (3.8), it should be noted that the term $T_{\mu\nu}^{[\mathrm{IMT}]}$ is the same effective matter obtained in the IMT for which GR was considered as an underlying theory. However, for the BD theory, due to the presence of the BD scalar field, an extra effective energy momentum tensor is also induced on the hypersurface:

$$T_{\mu\nu}^{[\phi]} \equiv \frac{\epsilon\overset{*}{\phi}}{2\psi^2}\left[\overset{*}{g}_{\mu\nu} + g_{\mu\nu}\left(\frac{\omega\overset{*}{\phi}}{\phi} - g^{\alpha\beta}\overset{*}{g}_{\alpha\beta}\right)\right]. \tag{3.9}$$

It is seen that $T_{\mu\nu}^{[\phi]}$ depends on the first derivatives of ϕ with respect to the fifth coordinate.

3. The wave equation associated with the MBDT:

$$\mathcal{D}^2\phi = \frac{1}{2\omega + 3}\left[8\pi T^{[\mathrm{MBDT}]} + \phi\frac{dV(\phi)}{d\phi} - 2V(\phi)\right], \tag{3.10}$$

where $V(\phi)$ is obtained from

$$\phi\frac{dV(\phi)}{d\phi} \equiv -2(\omega+1)\left[\frac{(\mathcal{D}_\alpha\psi)(\mathcal{D}^\alpha\phi)}{\psi} + \frac{\epsilon}{\psi^2}\left(\overset{**}{\phi} - \frac{\overset{*}{\psi}\overset{*}{\phi}}{\psi}\right)\right]$$

$$-\frac{\epsilon\omega\overset{*}{\phi}}{\psi^2}\left(\frac{\overset{*}{\phi}}{\phi} + g^{\mu\nu}\overset{*}{g}_{\mu\nu}\right) + \frac{\epsilon\phi}{4\psi^2}\left[\overset{*}{g}^{\alpha\beta}\overset{*}{g}_{\alpha\beta} + (g^{\alpha\beta}\overset{*}{g}_{\alpha\beta})^2\right]. \tag{3.11}$$

4. An effective equation that is counterpart to the conservation equation introduced in the IMT:

$$G_{\alpha 4}^{(5)} = \psi\mathcal{D}_\beta P^\beta{}_\alpha = \frac{\omega\overset{*}{\phi}(\mathcal{D}_\alpha\phi)}{\phi^2} + \frac{\mathcal{D}_\alpha\overset{*}{\phi}}{\phi} - \frac{\overset{*}{g}_{\alpha\lambda}(\mathcal{D}^\lambda\phi)}{2\phi} - \frac{\overset{*}{\phi}(\mathcal{D}_\alpha\psi)}{\phi\psi}. \tag{3.12}$$

It is worthy noting that equations (3.7) and (3.10) can be retrieved from the action

$$\mathcal{S}_{\mathrm{BD}}^{(5)} = \int d^4x\sqrt{-g}\left[\phi R^{(4)} - \frac{\omega}{\phi}g^{\alpha\beta}(\mathcal{D}_\alpha\phi)(\mathcal{D}_\beta\phi) - V(\phi) + 16\pi L_{\mathrm{matt}}^{(4)}\right], \tag{3.13}$$

where $\sqrt{-g}T_{\mu\nu}^{[\mathrm{MBDT}]} \equiv 2\delta\left(\int d^4x\sqrt{-g}\, L_{\mathrm{matt}}^{(4)}\right)/\delta g^{\alpha\beta}$.

In summary, using (2.2) as the background line element applying a suitable reduction method, the effective MBDT field equations, i.e., (3.6), (3.7), (3.10) and (3.12), are obtained from equations (3.2) and (3.5). It is worth noting that both $T_{\mu\nu}^{[\mathrm{MBDT}]}$ and $V(\phi)$ can be considered as fundamental quantities rather than ad hoc phenomenological assumptions taken in the conventional generalized BD frameworks.

4 Exact BD-KS anisotropic vacuum solutions in five-dimensions

In this section, let us review the solutions of the $5D$ BD field equations (3.2) and (3.5) for the extended anisotropic Kantowski-Sachs metric [12], i.e.,

$$dS^2 = -dt^2 + a^2(t)dr^2 + b^2(t)\left(d\theta^2 + \sin^2\theta d\phi^2\right) + \epsilon\psi^2(t)dl^2, \quad (4.1)$$

where t is the cosmic time; $a(t)$, $b(t)$ and $\psi(t)$ are cosmological scale factors.

Assuming that the BD scalar field depends only on time, i.e., $\phi = \phi(t)$, and using equations (3.2), (3.5) and (4.1), we obtain

$$\frac{\ddot{\phi}}{\phi} + \frac{\dot{\phi}}{\phi}\left(\frac{\dot{a}}{a} + \frac{2\dot{b}}{b} + \frac{\dot{\psi}}{\psi}\right) = 0, \tag{4.2}$$

$$\frac{\dot{b}}{b}\left(\frac{2\dot{a}}{a} + \frac{\dot{b}}{b}\right) + \frac{\dot{\phi}}{\phi}\left[\frac{\dot{a}}{a} + \frac{2\dot{b}}{b} - \frac{\omega}{2}\left(\frac{\dot{\phi}}{\phi}\right) + \frac{\dot{\psi}}{\psi}\right] + \frac{\dot{\psi}}{\psi}\left(\frac{\dot{a}}{a} + \frac{2\dot{b}}{b}\right) + \frac{1}{b^2} = 0, \tag{4.3}$$

$$\frac{\ddot{b}}{b} + \frac{\dot{b}}{b}\left(\frac{\dot{a}}{a} + \frac{\dot{b}}{b} + \frac{\dot{\phi}}{\phi}\right) + \frac{1}{2}\left[\frac{\ddot{\psi}}{\psi} + \frac{\dot{\psi}}{\psi}\left(\frac{\dot{a}}{a} + \frac{4\dot{b}}{b} + \frac{\dot{\phi}}{\phi}\right)\right] + \frac{1}{b^2} = 0, \tag{4.4}$$

$$\frac{\ddot{a}}{a} + \frac{\dot{a}}{a}\left(\frac{2\dot{b}}{b} + \frac{\dot{\phi}}{\phi}\right) + \frac{1}{2}\left[\frac{\ddot{\psi}}{\psi} + \frac{\dot{\psi}}{\psi}\left(\frac{3\dot{a}}{a} + \frac{2\dot{b}}{b} + \frac{\dot{\phi}}{\phi}\right)\right] = 0, \tag{4.5}$$

$$\frac{\ddot{a}}{a} + \frac{2\ddot{b}}{b} + \frac{\dot{b}}{b}\left(\frac{2\dot{a}}{a} + \frac{\dot{b}}{b}\right) + \frac{\dot{\phi}}{\phi}\left[\frac{\omega}{2}\left(\frac{\dot{\phi}}{\phi}\right) - \frac{\dot{\psi}}{\psi}\right] + \frac{1}{b^2} = 0, \tag{4.6}$$

where an overdot denotes a derivative with respect to t.

Among (4.2)-(4.6), only four equations are independent, where we also have four unknowns a, b, ϕ and ψ. To solve these equations we introduce a new time coordinate η related to t as

$$dt = bd\eta. \tag{4.7}$$

After some manipulation, it is easy to show that equations (4.2)-

(4.6) can be rewritten as

$$ab\psi\phi' \; = \; c_1, \tag{4.8}$$

$$Z'' + Z \; = \; 0, \qquad \text{where} \qquad Z \equiv ab\phi\psi, \tag{4.9}$$

$$(XZ)' + 2Z \; = \; 0, \qquad \text{where} \qquad X \equiv [\ln(ab^2)]', \tag{4.10}$$

$$[\ln(YZ)]' \; = \; 0, \qquad \text{where} \qquad Y \equiv [\ln(a\psi^{\frac{1}{2}})]', \tag{4.11}$$

$$\frac{b'}{b}\left(\frac{2a'}{a}+\frac{b'}{b}\right) + \frac{\phi'}{\phi}\left[\frac{\psi'}{\psi}-\frac{\omega}{2}\left(\frac{\phi'}{\phi}\right)\right] \tag{4.12}$$
$$+ \; \left(\frac{a'}{a}+\frac{2b'}{b}\right)\left(\frac{\phi'}{\phi}+\frac{\psi'}{\psi}\right) + 1 = 0,$$

where a prime stands for $d/d\eta$ and c_1 a constant of integration.

A solution of the equation (4.9) is $Z(\eta) = Z_0 \sin(\eta + \eta_0)$ where $Z_0 \neq 0$ is an integration constant. It should be noted that η_0 can be set equal to zero without loss of generality. Therefore we get the following exact solutions

$$a(\eta) \; = \; a_0\left[tan\left(\frac{\eta}{2}\right)\right]^{m_1}, \qquad b(\eta) = b_0 sin\eta\left[tan\left(\frac{\eta}{2}\right)\right]^{m_2}, \tag{4.13}$$

$$\phi(\eta) \; = \; \phi_0\left[tan\left(\frac{\eta}{2}\right)\right]^{m_3}, \qquad \psi(\eta) = \psi_0\left[tan\left(\frac{\eta}{2}\right)\right]^{m_4}, \tag{4.14}$$

where m_i (with $i = 1, 2, 3, 4$) were defined as

$$m_1 \; \equiv \; \frac{2}{3}\left(2\alpha+\beta\right), \qquad\qquad m_2 \equiv -\frac{1}{3}\left(2\alpha+\beta\right), \tag{4.15}$$

$$m_3 \; \equiv \; \beta, \qquad\qquad m_4 \equiv -\frac{2}{3}\left(\alpha+2\beta\right). \tag{4.16}$$

In equations (4.13)-(4.16), a_0, b_0, ϕ_0, ψ_0, α and $\beta \equiv \frac{c}{Z_0}$ are either integration constants or parameters of the model. Moreover, using (4.12), we have easily shown that $a_0 b_0 \phi_0 \psi_0 = Z_0$ and

$$4\alpha^2 \; + \; 6\alpha\beta + \left(\frac{3\omega}{2}+5\right)\beta^2 - 3 = 0, \tag{4.17}$$

which can be rewritten as

$$\omega = -\frac{2(4\alpha^2+6\alpha\beta+5\beta^2-3)}{3\beta^2}. \tag{4.18}$$

Furthermore, we obtained the following constraints

$$\sum_{i=1}^{4} m_i = 0, \qquad \sum_{i=1}^{4} m_i^2 = 2 - \omega m_3^2, \qquad (4.19)$$

where we have used (4.18).

Let us make some remarks about these solutions:

- We should note that there are only two independent parameters, so the third one is constrained by (4.17).

- Assuming $\alpha = -2\beta$, so $\psi(\eta)$ takes constant values. Therefore, our herein solutions can be reduced to those found for a vacuum $4D$ space-time in the context of the BD cosmology.

- If β tends to zero, then $|\omega|$ tends to infinity. In this particular case, the BD scalar field takes constant values. Consequently, the solutions (4.13) and (4.14) can be reduced to those found in an empty $5D$ space-time in GR. (It is worth noting that when the BD coupling parameter goes to infinity, the BD solutions may reduce (but not always [16, 17]) to their GR counterparts).

5 Effective BD-KS cosmology on a four dimensional hypersurface

In this section, we give a summary of the reduced BD-KS cosmology on the hypersurface; for a more in-depth analysis, see [12].

Applying the framework reviewed in section 3, we will obtain the components of the induced EMT and the scalar potential. Then, we will proceed to obtain the solutions associated with the MBDT cosmology.

Using (3.8) and (4.1), we can easily obtain the non-vanishing components of the induced EMT in terms of the comoving time:

$$\frac{8\pi}{\phi} T^{0[\text{MBDT}]}_{0} = -\frac{\ddot{\psi}}{\psi} + \frac{V(\phi)}{2\phi}, \qquad (5.1)$$

$$\frac{8\pi}{\phi} T^{1[\text{MBDT}]}_{1} = -\frac{\dot{a}\dot{\psi}}{a\psi} + \frac{V(\phi)}{2\phi}, \qquad (5.2)$$

where replacing a by b in relation (5.2), we get the other component $\frac{8\pi}{\phi} T^{2[\text{MBDT}]}_{2}$ (that is equal to $\frac{8\pi}{\phi} T^{3[\text{MBDT}]}_{3}$). Moreover, to obtain $V(\phi)$ we will use (3.11).

It is also easy to obtain the energy density ρ and pressures P_i (where $i = 1, 2, 3$) in terms of η:

$$\rho(\eta) \equiv -T_0^{0[\text{MBDT}]} = \frac{\phi(\eta)}{8\pi b^2(\eta)}\left(\frac{\psi''}{\psi} - \frac{b'\psi'}{b\psi}\right) - \frac{V(\eta)}{16\pi}, \quad (5.3)$$

$$P_1(\eta) \equiv T_1^{1[\text{MBDT}]} = -\frac{\phi(\eta)}{8\pi b^2(\eta)}\frac{a'}{a}\frac{\psi'}{\psi} + \frac{V(\eta)}{16\pi}, \quad (5.4)$$

$$P_2(\eta) \equiv T_2^{2[\text{MBDT}]} = P_3(\eta) \equiv T_3^{3[\text{BD}]} = \quad (5.5)$$
$$-\frac{\phi(\eta)}{8\pi b^2(\eta)}\frac{b'}{b}\frac{\psi'}{\psi} + \frac{V(\eta)}{16\pi}.$$

Moreover, equation (3.11) for our model reduces to

$$\frac{dV(\phi)}{d\phi} = \frac{2(1+\omega)}{b^2(\eta)}\left(\frac{\phi'}{\phi}\right)\left(\frac{\psi'}{\psi}\right). \quad (5.6)$$

To get the energy density and pressures, we should first obtain the induced potential. Substituting solutions (4.13) and (4.14) into equation (3.11), we get:

$$V(\eta) = \frac{V_0}{2}\int du\,(1+u^2)^4\,u^m, \quad (5.7)$$

where

$$m \equiv \frac{1}{3}(4\alpha + 5\beta - 15), \quad V_0 \equiv -\frac{(1+\omega)(\alpha + 2\beta)\beta^2\phi_0}{12b_0^2}, \quad (5.8)$$

$$u(\eta) \equiv \tan\left(\frac{\eta}{2}\right). \quad (5.9)$$

Equation (5.7) gives

$$V(\eta) = V_0 u^m \sum_{n=0}^{4}\left[\binom{4}{n}\frac{u^{2n+1}}{m+(2n+1)}\right], \quad (5.10)$$

where, without loss of generality, the integration constant has been set equal to zero.

Before proceeding, let us outline some particular but important cases:

- Assuming $\beta = 0$, we get $\phi = \phi_0 = $ constant. For this particular case, the solutions obtained in the previous section reduce to the corresponding KS cosmological model in GR in five dimensions. Therefore, our herein model may reduce to of the KS model in the context of the IMT.

- For the particular case where $\omega = -1$, we can get close resemblance between the scalar–tensor theories and supergravity [18].

- Assuming $\alpha = -2\beta$, $\psi(\eta)$ takes constant values, hence all the solutions we obtained up to now reduce to the corresponding ones retrieved for the standard BD theory in four dimensions.

Let us return to the general case. To obtain the energy density and pressures, we substitute $a(\eta)$, $b(\eta)$, $\phi(\eta)$ and $\psi(\eta)$ from relations (4.13) and (4.14) into (5.3)-(5.5). Therefore, we obtain the components of the induced EMT in terms of η

$$\rho(\eta) = T_0 \left\{ \left[(\beta + 2) + (\beta - 2)u^2 \right] \left(1 + u^2 \right)^3 \right. \tag{5.11}$$
$$\left. + (1 + \omega)\beta^2 \sum_{n=0}^{4} \binom{4}{n} \frac{u^{2n+1}}{m + (2n+1)} \right\} u^m,$$

$$P_1(\eta) = T_0 \left\{ \left[\frac{2}{3}(2\alpha + \beta) \right] u \left(1 + u^2 \right)^4 \right. \tag{5.12}$$
$$\left. - (1 + \omega)\beta^2 \sum_{n=0}^{4} \binom{4}{n} \frac{u^{2n+1}}{m + (2n+1)} \right\} u^m,$$

$$P_2(\eta) = \frac{T_0}{3} \left[(3 - 2\alpha - \beta) - (3 + 2\alpha + \beta)u^2 \right] u^{m+1} \left(1 + u^2 \right)^3$$
$$- T_0 (1 + \omega)\beta^2 u^m \sum_{n=0}^{4} \binom{4}{n} \frac{u^{2n+1}}{m + (2n+1)}, \tag{5.13}$$

where

$$T_0 \equiv \frac{\phi_0(\alpha + 2\beta)}{192\pi b_0^2}, \tag{5.14}$$

and we have used (5.10).

We should note that as $T^{1[\text{MBDT}]}_1 \neq T^{2[\text{MBDT}]}_2$, therefore, the induced matter is not a perfect fluid. In the noncompactified KK frameworks described above, it has been shown that the induced EMT obeys the conservation law that is for our herein model is given by

$$\dot{\rho} + \sum_{i=1}^{3} (\rho + P_i) H_i = 0, \tag{5.15}$$

where $H_1 = \dot{a}/a$ and $H_2 = H_3 = \dot{b}/b$ denote the directional Hubble parameters. Using (4.13) and (4.14), equation (5.15), in terms of the new time coordinate, is rewritten as

$$u\rho'(\eta) \ + \ \frac{1}{3}(2\alpha + \beta)\left(1 + \zeta u^2\right)\left[P_1(\eta) - P_2(\eta)\right] \tag{5.16}$$
$$+ \ \left(1 - \zeta u^2\right)\left[\rho(\eta) + P_2(\eta)\right] = 0,$$

where $u(\eta)$ is given by (5.9). Substituting the energy density and the pressures from (5.11)-(5.13) into (5.16) and then employing equation (4.17), it has been shown that (5.16) is satisfied for for our herein both BD-KS model [43].

In what follows, let us study the properties of some physical quantities such as the average Hubble parameter H, the deceleration parameter q, the spatial volume V_s, mean anisotropy parameter A_h, and the expansions for scalar expansion θ and the shear scalar σ^2:

$$V_s \ = \ A^3(t) = a(t)b^2(t), \qquad \theta = 3H = \left(\frac{\dot{a}}{a} + \frac{2\dot{b}}{b}\right),$$

$$A_h \ = \ \frac{1}{3}\sum_{i=1}^{3}\left(\frac{\Delta H_i}{H}\right)^2, \qquad \text{where} \quad \Delta H_i = H_i - H,$$

$$q \ = \ \frac{d}{dt}\left(\frac{1}{H}\right) - 1 = -\frac{A\ddot{A}}{\dot{A}^2},$$

$$\sigma^2 \ = \ \frac{1}{2}\sigma_{ij}\sigma^{ij} = \frac{1}{3}\left(H_1^2 + H_2^2 - 2H_1 H_2\right), \tag{5.17}$$

where $i, j = 1, 2, 3$ and A(t) denotes the mean scale factor of the universe. By substituting $a(\eta)$, $b(\eta)$, $\phi(\eta)$ and $\psi(\eta)$ from relations (4.13) and (4.14) into (5.17), we get

$$V_s(\eta) = A^3(\eta) = a_0 b_0^2 \left(\frac{2u}{1+u^2}\right)^2,$$

$$\theta(\eta) = 3H(\eta) = \left(\frac{1-u^4}{2b_0 u^2}\right) u^{\frac{1}{3}(2\alpha+\beta)},$$

$$A_h(\eta) = (2\alpha+\beta)\left(\frac{1+u^2}{1-u^2}\right)\left[\left(\frac{2\alpha+\beta}{2}\right)\left(\frac{1+u^2}{1-u^2}\right)-1\right]+\frac{1}{2},$$

$$q(\eta) = \frac{(4+2\alpha+\beta)u^4+4(1+u^2)-(2\alpha+\beta)}{2(1-u^2)^2},$$

$$\sigma^2(\eta) = \frac{1}{3b_0^2}\left(\frac{1+u^2}{2u}\right)^4\left[(2\alpha+\beta)-\left(\frac{1-u^2}{1+u^2}\right)\right]^2 u^{\frac{2}{3}(2\alpha+\beta)}. \quad (5.18)$$

It is worthy to note that when ω goes to infinity, our herein BD-KS solutions may reduce to the corresponding ones in IMT. For instance, as seen from relation (4.18), when β tends to zero, then $|\omega|$ goes to infinity. Therefore, for this particular case, we get $\phi = \phi_0 = $ constant and

$$\alpha = \pm\frac{\sqrt{3}}{2}, \quad (5.19)$$

where we have used (4.17). In this case, setting $\beta = 0$ and substituting α from (5.19) into (4.13) and (4.14), we obtain

$$a(\eta) = a_0 u^{\pm\frac{2\sqrt{3}}{3}}, \quad b(\eta) = b_0 \sin\eta\, u^{\mp\frac{\sqrt{3}}{3}}, \quad \psi(\eta) = \psi_0 u^{\mp\frac{\sqrt{3}}{3}}, \quad (5.20)$$

where $u = u(\eta)$ was given by (5.9). Obviously, equations (4.9)-(4.12) are satisfied by substituting these solutions, where assuming $c_1 = 0$, it is seen that (4.8) yields an identity, $0 = 0$. Moreover, for this particular case, the induced scalar potential vanishes, see (5.7). Therefore, employing (5.11)-(5.13), we obtain

$$\rho(\eta) = 2T_0\left(1-\zeta u^2\right)\left(1+\zeta u^2\right)^3 u^{m+1}, \quad (5.21)$$

$$P_1(\eta) = \pm\frac{2\sqrt{3}}{3}T_0\left(1+\zeta u^2\right)^4 u^{m+1}, \quad (5.22)$$

$$P_2(\eta) = \frac{T_0}{3}\left[(3\mp\sqrt{3})-3(3\pm\sqrt{3})\zeta u^2\right]\left(1+\zeta u^2\right)^3 u^{m+1}; \quad (5.23)$$

216

where

$$T_0 \equiv \frac{\pm\sqrt{3}\phi_0}{384\pi b_0^2}, \qquad m = \frac{1}{3}(\pm 2\sqrt{3} - 15). \qquad (5.24)$$

Finally, we should note that for our herein anisotropic model, retrieving the analytical solutions in terms of the cosmic time is a complicated procedure.

6 Conclusions and discussions

The development of KK theory has a lengthy and significant history. The cylinder condition is one of the fundamental flaws in Kaluza's primary mechanism. Moreover, the compactness condition of the Klein's version has produced certain physical problems [19]. Many attempts have been made to address the problems with the original KK theory by modifying their underlying assumptions. One of the more intriguing contemporary versions of the original KK theory is the IMT, which avoids recurring issues by using a completely general form for the underlying metric. We have provided the main formalism of the IMT in Section 2 of this review. The interpretation of the field equations associated with the IMT and their applications in cosmology and astrophysics have been extensively presented in the literature, see for instance, [1, 4, 20, 21, 22, 23, 24, 25, 26].

The MBDT [11] and MSBT [8, 9, 27] are two significant instances of recently established generalized IMT versions. The underlying theories for these updated frameworks replace GR with the BD and SB theories. Higher-dimensional matter fields were also taken into account when developing these modified theories, resulting in more generalized frameworks. In the MBDT and MSBT contexts, not only matter fields but also an induced scalar potential emerge as a result of the existence and curvature of the extra dimension, which distinguishes these theories from their standard counterparts. To be more specific, let us concentrate on the BD theory. In the generalized versions of the standard BD theory, several ad hoc assumptions (such as varying *omega* [28] and manually adding potential to the action [29, 30]) have been made in order to depict evolution of our 4D universe, specifically for attaining accelerating scale factor [31, 32, 33]. However, in the MBDT, the fundamental induced potential as well as induced matter play the necessary role [11]. Therefore, for the cosmological models in the contexts of the MBDT and MSBT we do not need any phenomenological assumptions.

Aside from the FLRW model [34], anisotropic cosmological models have been studied in the context of GR or the alternative theories to GR (in 4D as well as higher-dimensional contexts) [35, 36, 37, 38, 39, 40], and noncompactified KK theories [41, 43, 42, 12]. In this paper, we presented the anisotropic BD-KS model as an application of the MBDT.

Last but not least, using the MSBT and MBDT, due to their enormous potential, in future studies to investigate a number of open problems will be beneficial.

Acknowledgments

I would like to thank the organizers of *'Sixth International Conference on the Nature and Ontology of Spacetime'*. I also acknowledges the FCT grants UID-B-MAT/00212/2020 and UID-P-MAT/00212/2020 at CMA-UBI plus the COST Action CA18108 (Quantum gravity phenomenology in the multi–messenger approach).

References

[1] J.M. Overduin and P.S. Wesson, *Phys. Rep.* **283**, 303 (1997).

[2] L. L. Williams, "Field Equations and Lagrangian for the Kaluza Metric Evaluated with Tensor Algebra Software", *Journal of Gravity* **2015**, 901870, doi:10.1155/2015/901870.

[3] P.S. Wesson and J. Ponce de Leon, *J. Math. Phys.* **33**, 3883 (1992).

[4] Wesson, P. *Space-Time-Matter: Modern Kaluza-Klein Theory*; World Scientific: Singapore, 1999.

[5] M.M. Lapola, P.H.R.S. Moraes, W. de Paula, J.F. Jesus, R. Valentim d and M. Malheiro Induced equation of state for the universe epochs constrained by the hubble parameter. *Chinese J. of Phys.* **2021**, *72*, 159.

[6] Doroud, N.; Rasouli, S.M.M.; Jalalzadeh, S. A class of cosmological solutions in induced matter theory with conformally flat bulk space. *Gen. Rel. Grav.* **2009**, *41*, 2637–2656.

[7] Rasouli, S.M.M.; Jalalzadeh, S. On the energy conditions in non-compact Kaluza-Klein gravity. *Ann. Phys.* **2010**, *19*, 276–280. https://doi.org/10.1002/andp.201010427.

[8] Rasouli, S.M.M.; Vargas Moniz, P. Modified Saez–Ballester scalar–tensor theory from 5D space-time. *Class. Quant. Grav.* **2018**, *35*, 025004. https://doi.org/10.1088/1361-6382/aa9ad3.

[9] Rasouli, S.M.M.; Pacheco, R.; Sakellariadou, M.; Moniz, P.V. Late time cosmic acceleration in modified Sáez–Ballester theory. *Phys. Dark Univ.* **2020**, *27*, 100446. https://doi.org/10.1016/j.dark.2019.100446.

[10] Rasouli, S.M.M.; Sakellariadou, M.; Vargas Moniz, P. Geodesic deviation in Saez-Ballester theory. *Physics of the Dark Universe* **2022**, *37*, 101112.

[11] Rasouli, S.M.M.; Farhoudi, M.; Vargas Moniz, P. Modified Brans–Dicke theory in arbitrary dimensions. *Class. Quant. Grav.* **2014**, *31*, 115002. https://doi.org/10.1088/0264-9381/31/11/115002.

[12] Rasouli, S.M.M.; Vargas Moniz, P. Extended anisotropic models in noncompact Kaluza-Klein theory. *Class. Quant. Grav.* **2019**, *36*, 075010. https://doi.org/10.1088/1361-6382/ab0987.

[13] Amani, H.; Halpern, P. Energy conditions in a modified Brans-Dicke theory. *Gen. Rel. Grav.* **2022**, *54*, 64. https://doi.org/10.1007/ s10714-022-02950-3.

[14] Wesson, P.S.; Ponce de Leon, J. Kaluza–Klein equations, Einstein's equations, and an effective energy-momentum tensor. *J. Math. Phys.* **1992**, *33*, 3883–3887.

[15] Rasouli, S.M.M.; Shojai, F. Geodesic deviation equation in Brans–Dicke theory in arbitrary dimensions. *Phys. Dark Univ.* **2021**, *32*, 100781. https://doi.org/10.1016/j.dark.2021.100781.

[16] A. Barros and C. Romero, *Phys. Lett. A* **173**, 243 (1993).

[17] N. Banerjee and S. Sen, *Phys. Rev. D* **56**, 1334 (1997).

[18] V. Faraoni, *Cosmology in Scalar Tensor Gravity* (Dordrecht:Kluwer Academic, 2004).

[19] Mashhoon, B.; Wesson, P.; Liu, H.Y. Dynamics in Kaluza-Klein gravity and a fifth force. *Gen. Rel. Grav.* **1998**, *30*, 555–571. https://doi.org/10.1023/A:1018814123514.

[20] S. Rippl, C. Romero and R. Tavakol, *Class. Quant. Grav.* **12**, 2411 (1995).

[21] P.S. Wesson, *Int. J. Mod. Phys. D* **17**, 635 (2008).

[22] M. Israelit, *Gen. Rel. Grav.* **41**, 2847 (2009).

[23] J.M. Romero and M. Bellini, *Phys. Lett. B* **674**, 143 (2009).

[24] A. Bejancu, C. Calin and H.R. Farran,, *J. Math. Phys.* **53**, 122503 (2013).

[25] P.S. Wesson and J.M. Overduin, *Adv. High Energy Phys.* **53**, 214172 (2013).

[26] P.H.R.S. Moraes, *Eur. Phys. J. C.* **75**, 168 (2015)

[27] Seyed Meraj Mousavi Rasouli, Shahram Jalalzadeh and Paulo Moniz, *Universe* **8**, 431 (2022).

[28] N. Banerjee and D. Pavon, *Phys. Rev. D* **63**, 043504 (2001).

[29] S. Sen and T. R. Seshadri, *Int. J. Mod. Phys. D* **12**, 445 (2003).

[30] Medine Ildes and Metin Arik, Exact Cosmological Solutions in Modified Brans-Dicke Theory. *Int. J. of Mod. Phys. D* **2023**, *32*, 2250131 (2023).

[31] Rasouli, S.M.M.; Vargas Moniz, P. Noncommutative minisuperspace, gravity-driven acceleration, and kinetic inflation. *Phys. Rev. D* **2014**, *90*, 083533. https://doi.org/10.1103/PhysRevD.90.083533.

[32] Rasouli, S.M.M.; Vargas Moniz, P. Gravity-Driven Acceleration and Kinetic Inflation in Noncommutative Brans-Dicke Setting. *Odessa Astron. Pub.* **2016**, *29*, 19. https://doi.org/10.18524/1810-4215.2016.29.84956.

[33] Rasouli, S.M.M.; Marto, J.a.; Vargas Moniz, P. Kinetic inflation in deformed phase space Brans–Dicke cosmology. *Phys. Dark Univ.* **2019**, *24*, 100269. https://doi.org/10.1016/j.dark.2019.100269.

[34] Rasouli, S.M.M.; Vargas Moniz, P. Exact Cosmological Solutions in Modified Brans-Dicke Theory. *Class. Quant. Grav.* **2016**, *33*, 035006. https://doi.org/10.1088/0264-9381/33/3/035006.

[35] Diksha Trivedi and A. K. Bhabor, *Int. J. of Mathematics Trends and Technology* **2021**, *67*, 20 (2021). doi:10.14445/22315373/IJMTT-V67I2P504.

[36] Andronikos Paliathanasis, Bianchi I Spacetimes in Chiral–Quintom Theory. *Universe* **2022**, *8*, 503.

[37] Genly Leon, Esteban González, Samuel Lepe, Claudio Michea, Alfredo D. Millano, *Eur. Phys. J. C* **2021** *81*, 414 https://doi.org/10.1140/epjc/s10052-021-09185-7

[38] Andronikos Paliathanasis, Jackson Levi Said and and John D. Barrow *Phys. Rev. D* **2018** *97*, 414044008.

[39] Özgür Akarsu, Nihan Katirci, N. Özdemir and J. Alberto Vázquez, *Eur. Phys. J. C* **2020** *80*, 32.

[40] Özgür Akarsu, Tekin Dereli and N. Katırcı; *Journal of Physics: Conference Series* **2022**, *2191*, 012001. doi:10.1088/1742-6596/2191/1/012001.

[41] J. Ponce de Leon and P. S. Wesson, *EPL* **84**, 20007 (2008).

[42] Rasouli, S.M.M. Kasner Solution in Brans–Dicke Theory and Its Corresponding Reduced Cosmology. *Springer Proc. Math. Stat.* **2014**, *60*, 371–375. https://doi.org/10.1007/978-3-642-40157-2_55.

[43] Rasouli, S.M.M.; Farhoudi, M.; Sepangi, H.R. Anisotropic Cosmological Model in Modified Brans–Dicke Theory. *Class. Quant. Grav.* **2011**, *28*, 155004. https://doi.org/10.1088/0264-9381/28/15/155004.

www.ingramcontent.com/pod-product-compliance
Lightning Source LLC
Chambersburg PA
CBHW072302210326
41519CB00057B/2545